WHAT PEOF

THE MYTHOL

A book that's badly needed and could be revolutionary. Bateman doesn't just describe evolutionary theories but digs out their imaginative roots. He shows how essential it is to understand our myths properly since myths are not lies but powerful visions that can shape all our perceptions. This matters; read it!
Mary Midgley, Philosopher and Author of *Beast and Man* and *The Myths We Live By*

In The Mythology of Evolution, Chris Bateman explains why ID and other creationist proposals are not science. But Bateman also has little tolerance for atheists who use evolution to argue against religion. If you want a thoughtful assessment of where evolution stands as a scientific theory, read this book.
Francisco Ayala, University Professor and Donald Bren Professor of Biological Sciences at the University of California, Irvine

This readable and insightful work restores sanity and reasonableness to philosophical reflection on evolution. Bateman carefully separates what is illuminating from what is misleading in the various metaphors and myths that have dominated evolutionary discourse... Scientists, philosophers and other intellectually interested readers will profit immensely from reading and reflecting on this thoughtful book.
John F. Haught, Landegger Distinguished Professor of Theology at Georgetown University

This book explores the many ways that extra-scientific metaphors, or myths, both color and constrain our view of evolution. Bateman shows that alternative myths are equally consistent with the data, but present a very different image of the evolutionary process.

John O. Reiss, Professor of Zoology & Department Chair Evolutionary and Developmental Morphology, Humboldt State University

The Mythology
of Evolution

The Mythology of Evolution

Chris Bateman

Winchester, UK
Washington, USA

First published by Zero Books, 2012
Zero Books is an imprint of John Hunt Publishing Ltd., Laurel House, Station Approach,
Alresford, Hants, SO24 9JH, UK
office1@jhpbooks.net
www.johnhuntpublishing.com
www.zero-books.net

For distributor details and how to order please visit the 'Ordering' section on our website.

Text copyright: Chris Bateman 2011

ISBN: 978 1 78099 649 3

A CIP catalogue record for this book is available from the British Library.

Design: Stuart Davies

Printed and bound by CPI Group (UK) Ltd, Croydon, CR0 4YY

We operate a distinctive and ethical publishing philosophy in all
areas of our business, from our global network of authors to
production and worldwide distribution.

CONTENTS

Biographical Notes

Chris Bateman is an outsider philosopher, game designer and author. Graduating with a Masters degree in Artificial Intelligence/Cognitive Science, he has since pursued highly-acclaimed independent research into how and why people play games, and written extensively on the neurobiology of play. In 2009, he was invited to sit on the IEEE's Player Satisfaction Modeling task force, in recognition for his role in establishing this research domain.

He has a lifelong interest in mythology and religion, and has travelled the world studying religious practices and beliefs. He has taken part in everything from Native American sweat lodges to Pagan solstice celebrations, as well as visiting Buddhist and Shinto shrines in Japan, and witnessing traditional tribal religions in Africa whilst living in the Sahel Reserve near the Sahara desert.

Chris works in the digital entertainment industry as an expert in game design, narrative and player satisfaction modeling, and has worked on more than thirty digital game projects over the last fifteen years. As a game designer and writer, he is best known for the games *Discworld Noir* and *Ghost Master*, as well as the books *Game Writing: Narrative Skills for Videogames*, *21st Century Game Design* and *Beyond Game Design*.

His first book of philosophy, *Imaginary Games*, was published in 2011, and adapts Professor Kendall Walton's make-believe theory of representation to digital and other kinds of games. His blog *Only a Game* (http://onlyagame.typepad.com) deals with both philosophy and digital game theory, and contains a prolific array of articles, many of which have been featured elsewhere.

Other Books by Chris Bateman

Imaginary Games, Zero Books, 2011

Beyond Game Design: Nine Steps Towards Better Videogames (Editor), Cengage Learning, 2009

Game Writing: Narrative Skills for Videogames (Editor), Thompson Delmar Learning, 2006

21st Century Game Design (with Richard Boon), Charles River Media, 2005

Forthcoming from zer0 Books

Chaos Ethics (2013)

Preface

The philosopher David Hume, writing in the eighteenth century, was highly disappointed when his book, *A Treatise of Human Nature* (1739), did not produce so much as a "murmur among the zealots" upon its publication. Hume, who remained hostile to organized religion throughout his life but does not appear to have been an atheist as such, took a perverse delight in poking at the hornet nests of his time, principally found amidst conventional Christianity. Today, we still encounter many dogmatic enclaves of the kind Hume opposed among the followers of religion, but to find the kind of scholastic inflexibility as Hume's "zealots" we would have to look deep into the musty corners of the scientific establishment.

This is not the place to catalogue at length the difficulty that novel ideas have in breaking through the thick crust of the peer review process, although many examples abound. The challenges Bernd Heinrich faced getting reviewers to believe his data on the mental acuity of crows (mentioned in Bekoff and Pierce, 2009) is a typical case from the field of ethology, where a dogmatic faith in the idea of animals as essentially little more than machines has persisted from Descartes to today. I do not believe that these kinds of resistances are entirely negative – a reluctance to adjust established scientific models assists in providing solid foundations to enquiry. But it also blocks innovation and sustains an orthodoxy that is antithetical to criticism even – or especially – where it is most needed.

The process of scientific research and the formation, modification and replacement of empirical theories is a sadly ugly business. I found I had no taste for academic life after my postgraduate studies, simply because the bureaucracy of the academy exposed to me by my supervisor disabused me of any notion I may have been harboring about the purity and geniality of the community of scholars and scientists. This is not to impugn the scientists themselves – I have met some wonderful people through my interest in scientific research – but I must acknowledge that becoming an academic is not an easy road to travel. My decision to prostitute myself to industry was, in retrospect, an easy out, whatever misgivings I may hold about it.

I make these observations to put into context the critique you are about to read. It is not my intent to 'attack evolution', but it is certainly my intent to reproach those who might level such an accusation against me – the "zealots" of the modern world. As a matter of fact, I find evolutionary theories a fascinating, although incomplete, account of the deep history of life on our planet, but I see no reason whatsoever that I should expect other people to feel the same way about this subject. Nor can I detect any terrible cost in specific communities choosing not to teach evolution in their schools, if only because I can think of a dozen other things – including philosophy – that it might be more practical for our children to learn.

It seems to me that if one wishes more people to accept the perspective so ably presented by Charles Darwin in 1859, there could be no better place to start than by stripping (or at least exposing) the crust of mythological prejudices and metaphorical nonsense that have been presented by otherwise sensible scientists as gospel truth. Divested of these unhelpful trappings, even the more moderate creationists might be willing to reconsider their beliefs concerning the science of the origins of life, whether or not they might ultimately change their mind on the key issues.

In this regard, it is undoubtedly a requirement that I disclose some information about my scientific and religious beliefs in order that my potential biases can be understood clearly. Yes, I am pro-religion – in fact, I have five religions (Zen Buddhism, Sufism, Hinduism, Christianity and Discordianism), all of which influence my morality, my philosophy and my way of life. But I am also deeply pro-science – and indeed have studied degrees in

Physics with Astrophysics, Computing and Information Systems and Artificial Intelligence/Cognitive Science, not to mention publishing academic papers on neurobiology and other suitably chewy subjects. However you choose to interpret my disposition in light of this background, it is my suggestion that I can bring a certain unique balance to the topic this book tackles.

This is a book about the 'nature of life', that is, the beliefs that people hold about life on our planet, very few of which can be assessed using empirical methods. The issue of whether we are all fundamentally selfish, of whether life is dominated by competition, or whether the gene is the ultimate and only meaningful locus for metaphorical 'selection' are all perspectives as deeply mythological as the question of whether some divine force was involved in setting life in motion "In the Beginning". People are entitled to their beliefs on these issues – they are not entitled to pretend these perspectives are not beliefs.

As well as 'the nature of life', this book is also about 'the life of nature', that is, the story of the natural world from inception to present. The study of life on Earth is not solely the preserve of biological sciences (including the chemistry of life) but is also always an *historical* investigation – and the goal of this, as in any history, is not simply to identify a minimal set of theories with explanatory power. It will not do, for instance, to pretend that the alliance between free-living bacteria that lead to the complex cells that *all* of our bodies are comprised from can be dismissed as unimportant because the self-interest of the co-operating bacteria explains why it happened. It does no such thing. Histories are accounts of *what* happened, and are never wholly reducible to theories purporting to explain *why*. For the most part, such theories radically overreach their plausible bounds.

One of the assertions that emerges from this enquiry is that the neutrality of the scientific endeavor with respect to knowledge is not credibly under threat from the meddlesome forces of "organized ignorance" allied to 'Religion' (as if this term somehow denoted a unified field), even though there are indeed people who flatly reject specific theories and findings, but it is undermined by the premature certainty of those who fight under the banner of 'Science' (as if this term somehow denoted a unified field). The creation of an 'enemy of science' to declare war upon has polarized discussions on certain topics – particularly evolution – to the point that anyone who does not toe the orthodox party line in either camp risks being ostracized or censured by their peers. This behavior should run contrary to the core values of both scientific research and religious practice.

Furthermore, if we are to uphold our existing Human Rights agreements, we have to allow people the freedom of belief to make up their own minds. If people choose to reject belief in evolutionary theories or any other aspect of modern science, that is their right under the terms of legislation that most Western countries have ratified. We have no justification for demanding such people believe the same way as us, no matter how preposterous their viewpoint may seem. It could well be that our Human Rights agreements need revision – it's certainly the case, as Alain Badiou (2001) has suggested, that they will be the cause of great evil if we try to force these agreements onto other nations and peoples against their consent. But anyone who claims to honor the ethical directives collected under the banner of Human Rights cannot then insist that people *must* believe specific things – whether those beliefs are religious or scientific in origin.

Truths that are claimed to be so utterly self-evident as to be mandatory should be able to stand on their own two feet, and if they cannot then we should question their claims either to truth or to self-evidence. It was a principal criticism of the Christian Church in previous centuries that it insisted upon identifying one single account of what was true and then attempted to compel everyone to adopt this version of truth. The problem with this method did not lie in it being based upon an incorrect account of the truth (although even Christians these days recognize that it was) but in it being an unjust and contemptible practice. It will remain profoundly monstrous irrespective of the veracity of the account of truth that motivates it.

Although this is a book of philosophy (a field renowned for impenetrable vocabulary) about evolutionary theories (a subject with its own bizarre lexicon) I have tried to simplify the language wherever possible. Consequently, you will find little reference to certain words that are the mainstay of discussions of evolution. I have used 'gene variant' instead of allele, for instance, and have avoided as much as is possible the terms 'genotype' and 'phenotype', preferring to talk about 'gene patterns' and 'animals' instead. 'Phenotype' is an antiseptic term, the kind of scientific jargon intended to suggest a superior objectivity and clarity of thought, while often promoting a particular ideology. These choices in word usage are not only intended to make the book easier to read, they serve to focus the discussions on what we are actually talking about. (As a further aid, a glossary of terms used in the book is included at the back, and key terms are italicized as they are introduced).

I must acknowledge some shoulders that I have borrowed as footholds in tackling this subject. I am indebted to the work of Mary Midgley, one of Britain's truly great philosophers, and one for whom her majesty lays not so much in the iconoclasm of her ideas but in her commitment to presenting those ideas in language suitable for a wider readership. If there is one thing that philosophers as a whole are terrible at doing, it is in adapting their oft-valuable trains of thought for passengers who prefer to travel unencumbered by bulky intellectual baggage. She has been an inspiration to me, and a delightful correspondent.

I owe a debt of gratitude to a number of philosophers and researchers for assistance in compiling the manuscript, including Michael Ruse, whose *Darwin and Design* I read with great delight not long after writing my interpretation of Kant on evolution included in this book; Michael Lynch, whose work in molecular biology is truly impressive; Armand Leroi for some assistance in respect to the Cichlids of Lake Tanganyika; and John O. Reiss, whose unique perspective on contemporary evolution is not only refreshing, it is securely rooted in an understanding of the history of the field.

Additional thanks are due to many of the people who have contributed directly or indirectly to the development of this project, in particular Sam Kalman, for spotting a giant problem with my originally planned subtitle; Peter Crowther, for arguing the contrary position, and for continually sharpening my arguments by opposition; and Kelly Briggs-Waldrop for her invaluable observations regarding a typical US reader's vocabulary, which alas required me to cut some beautiful but obscure terms such as 'panacea' and 'accretion' from the final manuscript.

I would also like to thank the scientists whose work on evolution inspired my interest in the field. Firstly, to Lynn Margulis whose work and writings transformed my understanding of both life and evolution in ways that would be impossible to adequately document. Also, to Richard Lewontin, that rarest of things, a scientific materialist who is cognizant of the restraints this form of thought occasionally imports. Similar gratitude is owed to the late Stephen Jay Gould, whose writing not only inspired my interest in the details of evolutionary theories, but may also have been an influence in my gradual move from fiction to non-fiction. The clarity and readability of his essays set the bar for what I hope I might someday achieve.

Lastly to my beloved nemesis, Richard Dawkins, whose work explicating complex evolutionary ideas is frequently first rate. His problem, alas, is a reluctance to stay on topic and a tendency towards a philosophical obtuseness that varies from careless to wanton. This leads him all-to-frequently to indulge in wild myth-making even while he harshly scolds others who happen to tell different myths concerning evolution. This book simply would not have come about without my jaw dropping surprise at just how far outside of the scientific lines his early work decides to paint, and I find it hilariously ironic that he has positioned himself as a bastion of scientific rigor after so many injudicious excesses of his own.

I also must thank him for percolating within me the necessary ire to break my silence and write publically about religious issues. I think, perhaps, that I might have drifted off quietly into bored agnosticism were it not for his caustic crusade giving me so many wonderful reasons to "return to the fold", so to speak. How different my life would have been had I not felt it necessary to counter the manifest prejudice against the followers of religion that Dawkins so perfectly epitomizes! Just as Kant was to find Hume so inestimable in motivating him to pursue his incredible philosophical counterpoint, so I too owe a similar debt to this most wonderful modern example of Hume's "zealots".

1. Scientific Mythology

Imagining the Birds

Every summer, when the sky is clear and the sun shines down on my back yard, my wife and I sit outside and watch the swifts, darting about in the sky, chasing insects. The small, jet-black birds are only about as long as my hand, and while their sweeping crescent wings resemble a boomerang they fly like little fighter jets, zooming above me with an innate grace and beauty I can only admire. According to the ancient stories, both the swift and I were made by a great and mysterious divine force that afforded each of us gifts suited to our diverse pursuits. The swift's aerodynamic wing was tailor-made for agile aerial acrobatics, for diving at more than a hundred miles an hour as it chases down the tiny insects that are its prey. According to contemporary stories, the swift's wing is an *adaptation*, its design-like perfection a consequence of blind *natural selection* acting on diverse gene patterns, as those birds who performed better against their rivals prospered.

Both these stories are myths of nature, and both rely upon our imagination for their understanding. There are, I believe, important truths to be found in both of them – although the truth in the ancient stories is quite unlike the truth of the scientific metaphor of *selection* and adaptation. But both can also mislead if they are taken too literally or simply. Furthermore, the modern story is not as antithetical to the old account as is sometimes imagined – Charles Darwin's metaphor of selection and fitness emerged directly from an older theological perspective and was not originally intended to be a rival to it. The two different myths agree that the natural world is filled with creatures whose talents are matched to their circumstances. In so much as they seem to disagree, it is only in the details of the mechanism proposed and the subsequent implications that might be drawn from it. We can say that both accept a *metaphor of design*, even though the way this image is used is significantly different.

While I watch the swifts from Manchester in the United Kingdom, my wife and I have also lived in Tennessee in the United States, where I have watched another bird that is uncannily like it. The barn swallow has more-or-less the same shape and nature as the swift, although it parades around with a different coat of paint – it has a blue back, and an orange breast, far more colorful duds than the swift's plain tuxedo. You would think from looking at these birds, which live some three thousand miles away from one another, that the incredible similarities between them must be a result of them being close relatives. But in fact, the swift is more closely related to another Tennessee native, the hummingbird, than it is to the barn swallows, and the swallows are songbirds while the swifts never sing a note.

Many of those who study evolution explain the similarities between the two birds in terms of adaptation: since both chase insects on the wing, an activity requiring agility and rapidity, their bodies have evolved to be as aerodynamically efficient as possible for high-speed flight, hence the striking similarity in both shape and behavior. How satisfying you find this explanation depends on what you think of adaptation as an explanation for the nature of life: there are many for whom this way of looking at the features of animals is deeply and uniquely gratifying, and there are some who find this viewpoint offensive, although chiefly for reasons that have nothing to do with the science behind adaptation.

I find descriptions of biological features in terms of adaptation to be perfectly adequate up to a point, but the meaning of adaptation for me is not to be found in the sense of intellectual accomplishment some feel in science having 'figured it out'. Rather, the perfect tuning of these birds speaks to me of the incredible adaptability of nature – the refinement of their talents to their chosen lifestyle inspires me, but I do not see this as an achievement of natural selection. In fact, as will become clear, selection is just a fiction, a convenient metaphor that gestures at the underlying truth. For me, that truth can be found in the story that the ancestors of swifts and barn swallows were inventive enough to begin to prey upon insects, and as a result of a gradual accumulation of advantages their descendents are now perfectly suited to that way of life.

The swifts I watch flying about at breakneck speeds above my back garden have spent their winter in Africa, and vacation in the United Kingdom and northern Europe when the deserts of that continent become unpleasantly hot. Similarly, barn swallows spend their summers in the pleasant coolness of the United States and Canada before heading south to the Caribbean, Costa Rica, Panama and South America for the winter. Among those who study animal behavior, this behavior is called *migration*.

When I suggested it was a 'vacation', many scientists would contend that this was "only a metaphor", and might even chide me for making an anthropocentric or unscientific assertion. Certainly, there is something fictional about my use of 'vacation' to describe the behavior of migrating birds, but the idea is no less valuable for involving a bit of imagination – it conveys quickly and easily the relationship between the two territories in which the birds live. Furthermore, the life sciences are packed full of creative images no more empirically grounded than 'bird vacations', most of which are comfortably ignored or accepted as something close to factual. It is the purpose of this book to help bring some of these myths to light, namely the metaphors that are used to present ideas from evolutionary studies, and also to suggest alternative myths that might be preferable for one reason or another.

Migration is a convenient example for demonstrating the kind of problem that develops when scientists lose sight of the distinction between fact and metaphor, between an explanatory model and the mythic images used to present that model. In studying bird migration, most scientists had assumed (perhaps subconsciously) that an explanation could be put together in terms of adaptations, that there were genetic features in the DNA of the animals that alone could explain how and why they did what they did. Migration in this respect was not seen as any different from wings in terms of how it was to be explained.

Although the scientists studying the phenomena came up with a variety of relevant mechanisms – including the polarization of light, geomagnetism, and the position of the stars and the sun – they all fell radically short of a complete explanation of migration (Bowlin et al, 2010). Some quickly reached the conclusion that in terms of the natural tendency for the birds to travel between more or less the same regions each time they migrated, it was going to be necessary to recognize that they had some kind of mental map (Ketterson and Nolan Jr., 1990), but even then explanations were offered in terms of the relevant brain regions such as the hippocampus, which is involved in memory recall.

In seeking explanations for behavior in nature, some scientists turn all too quickly to explanations in terms of genes. They believe that everything that we see in life is best explained in terms of the chemical patterns found in DNA, and that evolution in particular can only be truly understood as changes in the distribution of genes. There is some scientific truth to this point of view, but it is all too easily swallowed up by the myth of *gene supremacy* that treats minds as an irrelevancy next to the explanatory power of genetic science.

Real creatures – like swifts, barn swallows, or you and I – make their way in the world by virtue of a mind, a confluence of experience, instinct, motives and intentions that interprets what is happening to us and makes decisions about what we will do next. Genes are certainly involved in the development of the biological features that make that mind possible, such as eyes and ears which look at and listen to the world, sensory cortices that process that information, the hippocampus which retrieves memories of previous experiences to help interpret the senses, and which later stores new memories for the future. But genes no more decide what animals do than the blueprint of a hotel determines what people do inside its rooms. The blueprint may make some things more likely than others – no-one is taking a shower in a hotel room with no bathroom, for instance – but it is the people who stay in the room who decide what they do there, not their genes, and not the blueprint of the hotel.

Imagine for a moment that you are a bird, migrating south for the winter. You set off for your migration because a particular mood seized you, a restlessness that comes about every year when the days get shorter. As you fly southwards, you pass places you remember flying over before, but you aren't finding your way by memory alone. In your vision, you can *see the magnetic field of the earth* (Heyers et al, 2007), and you use this to guide your direction of flight. You could not find where you were going by just this 'compass' alone, however, because a compass means nothing without a map. You have to guide the journey using your magnetic sense in concert with the plan of the landscape in your memory, you have to adjust for gusts of wind that blow you off course, and you have to stick together with other birds on the same journey as best you can.

Now having imagined this, who would you say that was navigating? The natural response is that it was *you* – you navigated using your visual impression of the magnetic field as your compass, and your memory and view of the landscape as your map. Someone who advocates gene supremacy disagrees with this conclusion. They say that although it may *seem* like you were doing the navigating *really* it was the bird's genes that were responsible for the navigating, the flying, and everything else beside, and they did this by creating a program that the bird executed, like programming a computer with software. Sure there were some variables – the program had to be shown where it was going the first time, for instance – but once the data was in place everything happened more or less automatically.

This perspective misleads not only in terms of pretending that animals (including humans) don't have minds that make decisions and have motives, but also because it obscures the very important but secondary role that genes have in influencing behavior. For instance, in the previous thought experiment you set off to migrate because a 'mood seized you'. This change in mood was related to certain combinations of genes in your DNA: once your body had used the patterns specified by certain genes to make particular chemicals, an emotional experience compelled you to migrate under the appropriate conditions. But it is the *bird* that migrates; the genes only create the circumstances, like the blueprint of the hotel creating the possibilities of what might happen in it. No-one would suggest that the hotel blueprint *causes* people to checkout (even though the checkout desk is a part of the design of the hotel), but some people *do* suggest that genes directly cause behavior.

The 'migration mood' of the bird is similar to the mood that seizes you, for instance, when you nearly collide with another car on the road – feelings of surprise and fear, or perhaps anger if the driver of the other car was acting recklessly, or if you are tired or irritable. A specific gene holds the pattern for the chemicals related to each of these emotions in your body, one for epinephrine (the chemical of excitement and fear) and one for norepinephrine (the chemical of anger), but these genes are only a blueprint for the chemical that your body uses to produce those feelings. There are various parts of your brain that assess the situation you experience on the road to trigger those responses (and genes were originally involved in 'building' these), but you and your mind are not only involved in the process, you can and do completely subvert the 'biological program' for your own purposes all the time.

One of the problems with contemporary science is that people have a tendency to trust scientists to report accurately the true account of a situation. In many cases, scientists who have conducted the relevant research are uniquely positioned to recount the facts of the matter. But in order to do so, they frequently have to find a way to present the story in comprehensible terms, and this requires a metaphor, an imaginative fiction. What's more, much of the scientific research that takes place is *motivated* by such metaphorical stories – what I am calling myths – yet although these may be scientific myths, they are not part of empirical science as such. Rather, they are metaphysical extensions of the science, and we need to appreciate their intimate relationship with the scientific endeavor if we are not to be seriously misled about the nature of life.

Myths and Metaphysics

The myths of evolution that this book explores can be understood as metaphors, imaginative fictions, or as metaphysical stories. *Metaphysics*, the philosophical exploration of that which cannot be tested or proven, lies beyond the recognized borders of science, but science never manages to be entirely free of metaphysics. The belief that it does, that there is a notion of 'scientific truth' in some absolute sense, is itself a metaphysical belief. Indeed, one of the most prevalent confusions about science in modern times is that 'scientific' should be taken to mean 'proven true by science', rather than 'conducted in a spirit of formal investigation' or 'inferred from empirical research'. Thus, models that have been rejected, such as phlogiston (which was believed to account for combustion before oxygen's role was recognized) or the luminiferous aether (which accounted for the movement of light before photons were conceptualized), cannot be called 'unscientific' without falling under the criticism raised by Thomas Kuhn (1962):

If these out-of-date beliefs are to be called myths, then myths can be produced by the same sort of methods and held for the same sorts of reason that now lead to scientific knowledge. If, on the other hand, they are to be called science, then science has included bodies of belief quite incompatible with the ones we hold today. Given the alternatives, the historian must choose the latter. Out-of-date theories are not in principle unscientific because they have been discarded.

In a similar vein, when I talk about 'myths of evolution' I am not necessarily accusing various ideas of being unscientific, I am talking about stories that are spun out of the scientific theories in circulation. We are comfortable calling something like phlogiston a myth, because we presume that myths are not true, but this is not what I mean when I invoke the term 'myth'. When I, for instance, call 'the selfish gene' a myth of evolution, I do not mean that what is termed 'the gene-centered view' is not a valid scientific perspective, but rather that the idea of a 'selfish gene' is an abstract metaphorical embellishment that puts a particular spin onto an otherwise neutral concept. This is what I mean by 'myth' in this context: a metaphorical image used to present the facts in a particular way, or (synonymously) a metaphysical story that expresses a particular interpretative bias. These myths can be criticized or replaced but they can never be entirely eliminated, since there is no science without mythology in this sense.

No aspect of modern scientific thinking crosses so unnoticed into metaphysical territory as people's beliefs about evolution. Myths abound when dealing with this subject! It is not intelligent design which is the only or chief culprit in this regard – it is not hard to spot that this is a metaphysical belief about a scientific topic – but rather the various stories that are spun out of the numerous competing models developed for understanding the putative processes of natural selection. Because these models are all incomplete, speculative and by-and-large untestable, they accumulate a rich scientific mythology which is then mistaken for knowledge and (even more embarrassingly) used as the basis of assertions that in many cases are largely indistinguishable from those conducted under a religious paradigm.

This foray into the myths of evolution is not entirely unprecedented. Writing in 1985, the British philosopher, Mary Midgley, observed that while scientists sometimes complain about symbols and symbolism infiltrating science, the idea of a scientific process immune to myth and metaphor "seems to be both psychologically and logically impossible". She states:

The theory of evolution is not just an inert piece of theoretical science. It is, and cannot help being, also a powerful folk-tale about human origins. Any such narrative must have symbolic force... If we ask 'by what myths do people today support themselves?' we shall often find that they do it by myths which they wrongly suppose to be part of science.

Other philosophers have advanced similar arguments. Michael Ruse (2010) suggests that "without metaphors we are blind – metaphorically at least". He admits to being inclined towards the idea that metaphors are necessary for science, but is unsure how we might actually *prove* such a claim, while freely accepting that they are used throughout the sciences, and that he "cannot imagine science without them". Scientific metaphor is vital for offering new perspectives and stimulating scientists to "think of old things in new ways". They may be literally false, but they "push us in new directions".

According to Ruse, metaphors such as "natural selection, continental drift, force, work, attraction, charm, genetic code, Oedipus complex" all help to "structure experience and to build our models of understanding" even though Nature doesn't literally select, and the genes are not actually written in code. These ideas have "indispensable" advantages for discovery and therefore "for all of the dangers seem here to stay." The reason a mythology of science is unavoidable is that the distinction between these myths and legitimate scientific metaphor is very nearly non-existent. We do not simply read facts from reality with no involvement in the process – our own ideas influence our understanding of the facts.

Of course, my choice of 'myth' as a synonym for 'metaphor' in the context of science may sound deliberately provocative, since often when we talk about myths we *do* mean to imply falsehood, or a distortion of the truth. As far as I'm concerned, there is always *some* truth to any myth – the question isn't whether it is true or false, but in what sense there is truth in the fiction. When we talk of the mythical unicorn, for instance, the truth is that travelers did sell the spiral tusks of narwhal as the horn of the unicorn, which is how unicorns ended up listed in medieval bestiaries. Even though the unicorn was never a living creature as far as we know, the myth of the unicorn still contained some truth, since there *was* an animal with a single horn – it was even a mammal, albeit one that swam in the sea rather than galloped over the land. The myth may have got the facts wrong, but it was a distortion of the truth, not a pure fabrication or falsehood.

In the context of myths as factual distortions, molecular biologist Michael Lynch (2007) has described the evolutionary beliefs of certain scientists literally as 'myths'. He accepts as unproblematic the view that there has been an increase in the complexity of organisms over the past three and a half billion years, but recognizes that since evolution deals with observations that seek to explain historical events despite radically incomplete information, evolution "attracts significantly more speculation than the average area of science." This has led to various confusions that Lynch seeks to dispel concerning the role of natural selection in explicating the development of complex organisms such as you, me and the swifts.

Of particular note to our topic, Lynch denies that natural selection is "a necessary or sufficient force" to explain the emergence of life as we know it. He accepts evolution as key to understanding the history of life, but rejects as mythical the idea that natural selection is the whole of that story. He is far from the only scientist to hold this view, and it has to be taken seriously – although it also has to be understood carefully. The claim is not that *evolution* is a myth, but that evolution *attracts* myths, it accumulates them, as does any scientific domain to some degree. If we want to understand the nature of life, we need to know which scientific metaphors are no longer doing their job. This kind of realignment of our understanding of evolutionary science is another aspect of the story I am telling about the nature of life.

In this book, I explore some of the most famous evolutionary myths, and offer in each case an alternative story that is equally compatible with current theories but entirely different in both meaning and implication. My purpose is to show that while a scientific theory may seem to be neutral and unbiased, this is never the case. There are always values embedded in any presentation of scientific thought, and when we move from the academy into the popular media the mythic dimension becomes so pronounced it threatens to obscure the otherwise valid observations at its root. As Katrin Wiegman (2004) has noted, scientific metaphors can be misused in two ways – they can obscure the scientific concept they are supposed to clarify, or they can awaken unintended associations that mislead public understanding of science.

Discussion of almost all of the myths under consideration tends to be dominated by partisan camps, each holding firmly entrenched beliefs. To some extent, this situation is inevitable – it is as impossible to exist without beliefs as it is to live without drawing breath. However, the flaws in what others believe cannot serve as endorsement for our own beliefs: the establishment of truth is not a sporting match in which one team wins and another loses. Rather, truth is glimpsed when an issue is viewed from many diverse perspectives, and even then we can never be sure that there is not some unseen angle as yet unrevealed. If we want to really understand the truth about any topic, we may first have to find a way to draw a line between discernible facts and inscrutable metaphysics.

Popper's Milestone

All this talk of metaphysics may sound daunting. Those who don't use the word as part of their vocabulary may be concerned that discussions will be swept into mumbo-jumbo and irrelevancies, while those more familiar with the word may worry that it's meaning has become so blurred and undermined as to make it impossible to use meaningfully. However, while the topic of evolution does become complex in many ways it will largely be scientific issues that give us the biggest problems, not philosophical ones. As for the meaning of 'metaphysics', I will use it in a very simple way, one based on the work of the Austrian-born British philosopher Sir Karl Popper, who in his efforts to explore the question 'what is science?' inadvertently laid down a metaphorical stone marker which stands on the borders between science and metaphysics, providing a convenient option for distinguishing simply between the two.

All explanations of metaphysics tend to stray into historical sidelines, but perhaps the simplest approach is to claim (unfairly!) that the realm of metaphysics includes all the abstract nonsense philosophers discuss with no direct relevance to everyday life – things such as universals, necessity and possibility, ultimate causes of everything and so forth. It's easy to understand why someone might have cause to oppose metaphysics as meaningless, or at least unhelpful, given this admittedly misleading definition.

Perhaps the first major opposition of this kind came in the early nineteenth century from Auguste Comte (1830), who founded what is termed *positivism*, and suggested that science passed through an early religious phase, then a metaphysical stage, and would ultimately emerge as a pure science, with all trace of metaphysics eliminated. This led in turn to a movement known as *logical positivism*, which grew out of the renowned Vienna Circle of the 1920's that Wittgenstein was at first associated with, and later went to great lengths to disavow.

Enamored with the successes of science in the previous century, the logical positivists proposed that propositions gain their meaning by some specification of the actual steps taken to determine their truth or falsehood (the so-called verification principle). This focus on truth values is a common property in skeptical thought, and represents in part the importance philosophers have tended to ascribe to logic as a means of mathematically relating notions of truth and falsehood. At the heart of the movement was a particular confidence in science as the guide to discovering the world, expressed concisely by Rudolf Carnap (1928): "When we say that scientific knowledge is not limited, we mean: *there is no question whose answer is in principle unattainable by science."*

Although the logical positivists are the bad guys in this particular story, I do not mean to suggest that nothing good came out of the movement. In fact, the dialogue that began in Vienna led to many developments that were crucial to the refinement of scientific methods in the twentieth century. The logical positivists also added a great many useful terms to the language. Otto Neurath (1931), for instance, one of the leading figures of the movement before he was forced to flee Austria on account of the Nazi occupation, coined the term *physicalism* to refer to the belief that only physical matter exists (sometimes referred to as 'materialism').

What motivated the logical positivists as a whole was the view that metaphysics was *meaningless* since everything within this domain was unverifiable, and only that which was capable of being confirmed as true could (they reasoned) be important. Popper thought differently. To his mind, the success of science did not lie in it being more verifiable than metaphysics, or, for that matter, ethics. Unlike many scientists, Popper took a decidedly cool view on induction (much as the eighteenth century philosopher David Hume had done) and observed that it wasn't possible to *confirm* a universal scientific theory as this would require absolutely complete knowledge. However, it was possible to *disprove* a universal theory – all you needed was to find some disconfirming evidence.

Suppose you have an infinitely large sock drawer, or one with so many socks within it that it is effectively infinite. Every day, you pull two socks from this drawer, and every day both are black. You might conclude (via induction) that the sock drawer contains only black socks. But suppose one day we reach in and pull out a white sock. This immediately invalidates the 'black sock theory'! In the same way, Popper reasoned that a scientific theory cannot be proved true, but it can be falsified by contradictory evidence. Popper advanced the idea that falsification was a more useful criterion to apply to science than the verification principle, since universal theories can be falsified but they can never be conclusively proved, as with the sock drawer thought experiment. He therefore proposed that falsification be used as a boundary condition for science, and consequently that anything that could not be falsified belonged to the domain of metaphysics (Popper, 1959).

While the logical positivists held that metaphysics was meaningless because nothing that abstract could ever be verified, Popper *never* suggested that because metaphysics could not be falsified it was without meaning. Rather, he recognized that metaphysical statements tend to imply beliefs, and therefore that anything in the realm of metaphysics is a matter for individual belief. Nothing could rationally force a change in such beliefs, at least in terms of definitely proving them false, but this was categorically not the same as claiming that such beliefs were meaningless. By the time of Popper's knighthood in 1965, logical positivism was widely recognized as having run its course, in no small part thanks to Popper's contribution. (However, it was to live on in the form of a 'logical empiricism' expounded by exiles from the Vienna Circle living in the United States – but that is another story).

Sadly for Popper, his position that falsification could be used as a boundary condition for science was quickly under attack. As we will shortly see, Thomas Kuhn and others observed that scientists do not abandon their theories in the light of contradictory evidence, which was a central assumption in Popper's account, and his student Paul Feyerabend (who had severe personal issues with his ex-mentor Popper) went on to demonstrate that there were no lasting boundary conditions to *any human endeavor*, never mind science (Feyerabend, 1975). Nonetheless, this in no way invalidates the importance of Popper's work: logical positivism risked a particularly vicious form of intellectual fascism and Popper's contribution to its demise is something to be commended and celebrated.

Hostility towards metaphysics, however, certainly did not end with logical positivism. As E.J. Lowe (1995) has suggested, not only do contemporary philosophers look down on metaphysics but scientists involved in popularizing scientific beliefs and findings "do not conceal their contempt for philosophy in general as well as metaphysics in particular." As Lowe suggests, this attitude is problematic – if metaphysics is unnecessary, why is there so much widespread disagreement over fundamental matters? This fact alone "demonstrates the need for critical and reflective metaphysical inquiry", and it would be better if this could be pursued without being swept away with dogmatism. To put it another way: since we do not agree, even on purportedly objective issues such as science, the need for further discussion of metaphysics is amply demonstrated.

While Feyerabend may have been correct in asserting that there are no lasting boundary conditions to human endeavor, this does not preclude the useful deployment of Popper's concept of falsification. After all, the absence of *lasting* boundary conditions does not mean we cannot choose to erect boundaries wherever we wish. Indeed, if we *agree* to a particular boundary it may stand (by custom if by no other means) for as long as we wish. This, after all, is the basis by which one country distinguishes itself from another. It is for this reason that I talk about *Popper's milestone*, as a kind of arbitrary but helpful way of distinguishing not so much a boundary to science, but rather the gateway to the vast, untestable territories of metaphysics.

One final observation in respect to Popper's views on metaphysics is particularly appropriate to the argument of this book. Popper (1976) famously remarked that "Darwinism is not a testable scientific theory but a metaphysical research program", noting:

And yet, the theory is invaluable. I do not see how, without it, our knowledge could have grown as it has done since Darwin. In trying to explain experiments with bacteria which become adapted to, say, penicillin, it is quite clear that we are greatly helped by the theory of natural selection. Although it is metaphysical, it sheds much light upon very concrete and very practical researches. It allows us to study adaptation to a new environment (such as a penicillin-infested environment) in a rational way: it suggests the existence of a mechanism of adaptation, and it allows us even to study in detail the mechanism at work.

Thus Popper acknowledged that metaphysics were a part of science and indeed that this metaphysical element of science was important to the practical aspects of research, both in terms of motivation and in terms of comprehension of the phenomena being studied. This recognition – seldom discussed by scientists – will be highly significant when we come to address some of the key questions in respect to scientific metaphysics.

Somewhere out in the metaphorical landscape of knowledge lies an almost insignificant stone marker reading 'Erected at the boundary of Falsification'. It was set down by Popper to mark the end of science and the beginning of metaphysics. I suggest that even if we do not wish to use it to mark the border of science, Popper's milestone can still be usefully interpreted as saying: 'beyond this point lies metaphysics!' Armed with this marker, we are better equipped to examine the relationship between science and its metaphysical entailments.

Rewriting Science

When I first went to university, it was to study astrophysics. I was enthusiastic about the subject, and eager to learn. My expectation was that the workings of the universe were going to be explicated to me, and that I was going to be shown experiments which underpinned the theories of nature that had been developed, and from which those theories could be derived. But in fact, the theories were taught as strictly factual, and only then were the experiments enacted. Students, having already learned the prevailing models, approached each experiment with the certain knowledge of the expected results.

When we were asked, for instance, to take measurements from which to calculate the gravitational constant of the universe, there was no doubt as to what the correct answer was expected to be. In practice, few if any students produced results that would yield an answer sufficiently close to that dictated by prior theory. In the face of conflicting experimental evidence, most students would attempt the experiment again, usually once again yielding results which were not of the kind expected. After a few failed attempts, many students then adjusted their data in order to more closely resemble the expected results of the experiment.

This experience greatly challenged my image of science. It was not that the theory in question was fundamentally in error – if we had, for instance, averaged all the data gathered in the lab by all the students, the mean result would probably have resembled the expected results. But what I observed in the physics laboratory were students of science faced with unexpected data and then, instead of reporting this honestly as my preconceptions of the scientific method demanded, changing their measurements to conform with the expected results. That this was the way to get the highest marks from the laboratory experiments is not in doubt, but if the central value of science is truth, the students were not learning this value – they were learning how to toe the line with existing theory.

Scientists, in general, are not taught philosophy of science, and as a result rarely question their abstract beliefs about the nature of the scientific process. As a result, science has developed a persistent mythology, central to which is the idea that science uncovers the truth and that as the theories and techniques of science develop it moves closer and closer to absolute truth.

The celebrated historian and philosopher of science Thomas Kuhn, in his seminal work *The Structure of Scientific Revolutions*, put forward one of the decisive criticisms of this view. He observed, from study of the development of European science, that scientists' commitment to particular theoretical frameworks acted as a barrier to seeing data in a different light – even when that data was entirely contradictory to the theory. Like the students in the physics lab I encountered, theory somehow trumped observation.

Kuhn's historical analysis resulted in a model of science that denied the conventional belief in science as a process of gradual accumulation. The idea that science progressively assimilates building blocks and advances in steady and discrete steps did not match up to the history. Instead, Kuhn saw periods of what he termed *normal science*, during which scientists worked to adapt their current theories to a range of experimental observations, and periods of *scientific revolution* – when the old theories came into crisis, and were eventually supplanted by new theories.

The view of scientific research proposed by Kuhn was that a particular field does not emerge within science until scientists working in the relevant area begin to develop a common framework. Kuhn calls this framework a *paradigm*, a term he uses ambiguously in a number of subtly different contexts. On the one hand, a paradigm describes the collection of symbolic generalizations, experimental methods and common assumptions shared by practitioners of a given scientific field. On the other, paradigms represent specific exemplars of scientific puzzle solving, including both experimental and theoretical results to problems. We can see a paradigm as a set of common beliefs that a group of scientists share. Despite the intense faith that is sometimes invested in science, these beliefs do not necessarily correspond to reality – rather, they provide a framework that enables the scientists to investigate reality in a particular way.

Kuhn suggests that during periods of normal science, scientists are mostly attempting to "force nature into the preformed and relatively inflexible box that the paradigm supplies." No paradigm explains all the facts with which it can be confronted, and the general process of science therefore works on the solutions to puzzles – problems in adapting the theoretical models of that paradigm to the description of reality. This commitment to a particular paradigm is essential to the scientific process in Kuhn's view. Normal science is only able to proceed on the predicate that scientists know "what the world is like."

Anomalous data – that which cannot be made to fit into the box provided by the paradigm – is usually ignored, or interpreted in a manner consistent with the paradigm's assumptions, or else the theoretical framework of the paradigm is adjusted to compensate for the discrepancy. Experienced scientists, having had great success with their paradigm, are unwilling to abandon it in the light of contradictory evidence. This can be seen as a necessary commitment, since to give up a particular paradigm without adopting a new one would be to abandon science entirely: there can be no science without a particular model by which the investigatory process can proceed.

Kuhn observes that new paradigms gradually replace old ones as a result of a growing view (usually among newcomers to the field, since these are least committed to the old beliefs) that the old methods are no longer able to guide the exploration of a particular area in which the previous paradigm used to lead the way. But the transition between paradigms rarely proceeds without difficulties – the commitment to the old paradigm remains strong, and it is difficult for rational discussion to take place between two individuals who are committed to different paradigms.

There is a tendency to believe that when one scientific theory replaces another, it has done so because of some knock-down blow – some experiment or fact settled the matter once and for all. This can happen: the Michelson-Morley experiment did make theories of a luminiferous ether untenable, for instance (Michelson and Morley, 1887). Often, however, the elder scientists are unwilling or unable to abandon the paradigm which has led to such success and progress during their time. This is only natural; as the physicist Max Planck (1949) observed: "a new scientific truth does not triumph by convincing its opponents and making them see the light, but rather because its opponents eventually die, and a new generation grows up that is familiar with it."

There is no objective means of resolving these disputes between competing explanations. Scientists must necessarily premise their own paradigm in order to argue in its favor, a circular process that can nonetheless be quite persuasive. Although arguments are usually couched in terms of the capacity of the competing theories to explain experimental facts, it is rarely possible to determine the better theory solely by comparison with the facts. All significant scientific theories fit the facts to some extent, but when two theories are competing with each other the issue is not how much each theory fits the facts, per se, but rather whether or not one theory fits the facts *better* than another.

Even this may not be the means in which the commitment to a paradigm is changed. What is required is a fundamental alteration to the way a particular scientific field is practiced, and for a paradigm-changing argument to be persuasive what must be offered is something more than correspondence to facts – as already mentioned, most scientific theories can be made to adequately explain the facts they are faced with. For example, the theory of combustion in the presence of oxygen caused chemists to abandon its predecessor, phlogiston theory, but at the time that the new theory was being worked out, phlogiston adequately explained the majority of the observed facts. Actually, many of these facts *presumed* phlogiston in their formulation. However, the new idea carried with it the promise of future progress, and it was faith in this possibility that helped drive this paradigm shift forward.

Kuhn insisted that his theory of paradigm shifts was intimately connected with the way metaphors are used in science, since each paradigm was based upon different imaginative fictions. Devising a new paradigm is largely a process of determining a new metaphor for that area of science, and innovation was thus as much about imagination as it was about mathematics or experiments. As the philosopher of science Mary Hesse (1974) recognized, science as we now practice it cannot proceed without its metaphors. The explicit content of scientific theories just isn't rich enough to motivate research, and we thus need scientific metaphors to drive forward investigations. The same point is made by linguist George Lakoff and philosopher Mark Johnson

(1980) in their discussion of the pervasiveness of metaphors in life.

Scientific metaphors, as Ruse puts the matter, are indispensible in part because they possess a heuristic function: when we think of the heart as a pump, it leads to questions about rates of flow and so forth that would not otherwise be suggested (Ruse, 2003b). The application of a well-chosen metaphor stimulates new ways of looking at the world such that "a new truth emerges, one that is metaphorical at first but clearly understandable, and moves toward the literal... the shock of the metaphor pushes you to further thought." The campaign of the logical positivists was expressly against this kind of imaginative activity in science, but their attempt to eject metaphor from scientific thought was misguided. Ruse suggests that their project is now widely recognized as having failed:

Today, the consensus is that, far from being a sign of weakness, metaphors are an essential part of thought – including scientific thought. They are a prime force in causing people to think in new, imaginative, and fertile ways. They are a key ingredient in the heuristic side of thought, one of the most prized virtues of great science. Metaphors do not provide answers so much as they raise questions... Without the metaphor, the science would grind to a halt, if indeed it even got started.

Thus by introducing fresh metaphors, a new paradigm is able to offer quite different entities to its older rival. Consider the view of the world presented by Newton versus that presented by Einstein. In Newtonian physics, discrete matter had a property, mass, which innately attracted other objects with mass by a process called gravity. This (metaphorical) attraction was at first seen as a flaw, but later accepted as the nature of gravity once Newtonian mechanics grew in stature. Compare Einstein's theories of Special and General Relativity, which saw mass as a form of energy, and gravitational attraction as the result of a (metaphorical) curvature in space caused by the presence of energy. These are very different models of reality, as can be seen from the kinds of metaphors used to teach the different theories: Newtonian physics is often explained by analogy to the collisions of billiard balls, while relativity is often explained by analogy to a ball being placed on a rubber sheet, with the sagging of the sheet being equivalent to the curvature of space-time. The respective fictions accompanying these scientific models involve entirely different entities, and even the elements they seem to have in common are not strictly equivalent. Newtonian mass does not in any way anticipate the idea of mass that results from Special Relativity – and thinking about billiard balls does not lead to thinking about rubber sheets.

Part of the mythology of science is to see changes like these (from an old successful theory to a new one) as progressing towards truth, and additionally to view the old theory not as having been disproved but as a special case of the new theory. This tendency is particularly prevalent in relation to the paradigm shift from Newtonian to Einsteinian physics, and scientists can indeed demonstrate how the Newtonian equations can be derived from the later theories. But this process is misleading. An abandoned theory can always be viewed as a special case of its successor, but to do so it's necessary to transform the original theory into the terms of the new one – and this is something that can *only* be done in hindsight.

This rewriting of history is endemic to science. The scientists of earlier ages are represented in textbooks and so forth as having worked on the same set of problems, and in accordance with the same guidelines, as contemporary science. This presents the scientific process as if it were cumulative, but Kuhn insists that this viewpoint is a fiction created by ignoring the crises that accompany scientific revolution. Earlier scientists worked on very different problems, because their models of the world led them to different puzzles to solve. Because the results of scientific research do not seem to depend upon the history of their development (that is, scientific theories do not depend upon their historical origins for their veracity) it seems acceptable to abstract over the details of their development. In doing so, the nature of the scientific endeavor becomes distorted to appear both linear and cumulative.

Furthermore, this view misrepresents the nature of the abandoned paradigms. There is a tendency to look at certain discarded theories as myths that have been disproved – but this perspective will not stand up to scrutiny. Kuhn observes that the old paradigms were derived through the same essential scientific process as the new paradigms, and from a historical perspective both the old and the new must be seen as scientific – else all scientific knowledge can be accused of being a myth. It is not that the old beliefs were false and the new ones true – a future paradigm shift may well render the current ideas false by this reasoning. Rather, the science of the past contains views and beliefs wholly incompatible with the models of present scientists, yet all such paradigms are still by their very nature *scientific*. They proceeded from data and measurements with the goal of providing explanations of those observations.

Kuhn challenges our preconceptions of science as evolving *towards* objective truth, and instead suggests we should understand scientific progress as evolution *from* the community's prior state of knowledge. There is no perfect conception of reality that science is travelling towards, and there is no ideal goal state that science will ultimately evolve into. Such ideals are anachronistic fallacies that do not match up with the history of science. The idea of goal-directed progress was abandoned in the adoption of Darwin's theories of natural selection, and biology was no longer seen as evolution *towards* something. Instead, we now see a process that moves *from* primitive beginnings into the steady yet intermittent appearance of more complex and specialized organisms (although some mythology does persist concerning the view of evolution as progress, as we will discover in the next chapter). Kuhn contends the same situation applies in science – the progress we observe is not advancing *towards* anything, but proceeds from primitive theoretical beginnings into the steady yet intermittent appearance of more complex and specialized paradigms.

Ultimately, scientific knowledge, like language, is an intrinsic property of a group of people. To understand that knowledge, it is necessary to understand the nature and characteristics of the groups that create and use this knowledge. Science is the name we give to the practices of scientists, who by dedication to an empirical view of the world gradually refine their ideas, and produce exceptional instruments which in turn allow for the creation of new technologies. Later scientific theories appear to show progress – they are better at solving puzzles in often very different environments to those of their predecessors – but it is a mythological view of science that sees science as truth, or evolving towards truth. Science evolves as scientists refine their perspectives, but this refinement is an adaptation to new conditions of knowledge, and not an inevitable march towards perfection.

The Myths of Evolution

What follows is a critique of the mythological aspects of evolutionary theory presented in the context of Popper and Kuhn's work, and also in connection with the work of several other philosophers, particularly Mary Midgley, who has done much to clarify confusions at the borders of the scientific endeavor. Much of the discussion concerns specific scientific metaphors intended to express aspects of the nature of life, and the problems that they present if we take them at face value.

Psychologist and biologist Nicholas Thompson (2000) has

suggested that a useful scientific metaphor should be clearly specified. Inspired by the philosophical work of Mary Hesse, he suggests we need to be clear which parts of a scientific metaphor are indispensable and which are dispensable, the latter being whatever "the theorist can fairly disclaim". When these distinctions are not clear, critics can attack a theory for its dispensable implications, and proponents can protect a theory mistakenly by confusing what is tangential with what is central to its metaphors. I am tempted to say that these kinds of problems result from a failure to distinguish the metaphor from the myth, but although I am sympathetic to Thompson's argument I do not believe this kind of untangling is ever as clean cut as he suggests.

Metaphors, as I have written about elsewhere (2011), are imaginary games that cannot be dissected in such a sterile fashion. They have a life of their own, and we would be wise to recognize that there is no way to draw the line such that the useful metaphor can be entirely separated from the myth. As the philosopher Stephen Yablo (1998) recognized, the boundaries between literal and metaphorical are "about as blurry as they could be". Examining an old argument between logical positivist Rudolf Carnap and W.O. Quine, both of whom were looking to clearly define scientific truth, Yablo is forced to conclude that any kind of program that expects to eject all metaphors from science is ultimately doomed, since "like the poor, metaphor will be with us always."

Scientific metaphor and mythology are inescapably fused, and if Hesse is right that we need metaphor to conduct science we would be wise to appreciate the role of scientific myths in our understanding of the nature of life. In this book, I identify a set of evolutionary myths all of which have a legitimate element of scientific metaphor that has been tainted by some unfortunate ideological baggage. For each I propose an alternative 'spin', a different way of approaching the same subject. I do not want to suggest that my spin is 'better' – this judgment depends upon what the myth is being used for – but I do want to suggest that anyone who can understand the meaning of *both* myths in each case will have a superior understanding of the philosophical and scientific issues than someone who understands one or neither of these myths.

The first two myths concern notions of destiny, beginning with the *ladder of progress*, which suggests that evolution tells a story of progression from older, simpler forms to newer, more advanced forms. There are a considerable number of unwarranted assumptions embedded in this metaphor, and although few scientists publishing on evolution today would fall prey of this particular myth it can still be found running rampant in science fiction. Against this story, I propose the alternative myth of a *chain of inheritance* that connects all life together into one historical process.

The second myth is perhaps the single most famous myth of evolution, Herbert Spencer's catchy but misleading phrase *survival of the fittest*. This metaphor relates to the political ideas of Thomas Malthus and later concepts of social Darwinism, and is caught in an uncomfortable tension between being true simply by virtue of its definition, and being incorrect. Whatever this myth is intended to gesture at might be better captured in the alternative myth of a *refinement of possibilities*.

An additional pair of myths concern the relative importance of the gene in understanding life, and in particular that infamous phrase *the selfish gene*. This third myth concerns the metaphor by which the gene-centered view propagated itself into the popular imagination, and the substantial distortion it suffered in the process. Although Richard Dawkins bears responsibility for many of the problems in this regard, his later books also hold the seeds for a far less misleading myth, namely that *advantages persist*.

The closely related fourth myth, *kin selection*, depends in part upon the problems implicit in the selfish gene as a means of encapsulating the gene-centered view. There are palpable scientific problems with the notion of kin selection that need to be considered, but ultimately the problem is not that the idea is vacant but only that it obscures another, more positive myth of evolution, that *co-operation is an advantage* and, similarly, that *trust is an advantage*.

The final two myths of evolution concern the appearance of design in nature, and of these the most famous is the fifth myth: *intelligent design*. The nature of this metaphor was already insightfully exposed more than two centuries ago by the philosopher Immanuel Kant, but both advocates of the Intelligent Design movement and their opponents seem blissfully unaware of the history of the topic. Exposing this subject to close scrutiny offers the alternative myth of the *metaphor of design*, which both scientists and creationists might benefit from understanding.

The sixth myth is closely related to the metaphor of design, namely *adaptationism* – the belief that explanations in terms of natural selection can fully explain the history and nature of life. As with many of the other myths of evolution, there is a valid scientific perspective that relates to this viewpoint, but it entails serious risks of misunderstanding contemporary scientific models for the evolution of life. One possible way of defending against the excesses of adaptationism is to adopt a more restrained myth, that of the *conditions for existence*, such that rather than a close-fitting adaptation to environment being a consequence of evolution, it is merely a necessary requirement for life as we encounter it.

I also want to draw attention to a seventh myth, this time of science rather than of evolution, per se, that of *science as truth*. We have already seen the basic problem with this approach in Kuhn's paradigm account of scientific history, but it is worth examining more closely if we wish to avoid the kind of mythological confusions that have accumulated around evolutionary theories. Against this stance on the scientific endeavor I propose the alternative myth of *truth from fiction*, and suggest that if we are to genuinely understand what is true, we may have to begin by understanding fiction.

2. Progress and Destiny

Imaginary Selection

Every day, I take my dog to the park so he can run around, chase tennis balls, and sniff the grass to find out who else has been around. He's a Labrador Retriever, the most popular breed in the United Kingdom where I live, and also in the United States, where I also spend a lot of my time. He stands a little shorter than the other Labs in my neighborhood, but his silky black fur and irrepressibly friendly temperament have made him a lot of friends, both human and canine. He knew how to swim the very first time I threw a stick in the ocean, and he has slightly webbed toes and an Otter-like tail that assist him when he's in the water. He can't resist enthusiastically chasing after anything I throw for him, especially plastic flying disks which most of the time he brings straight back to me so I can throw them again. My dog – indeed every dog and every dog breed – is the product of two very different kinds of *selection*, one literal and one metaphorical.

In the first place, my dog is a product of selective breeding over the last few hundred years. His ancestors some five hundred years ago were a kind of dog known as the St. John's Water Dog which came about on the Canadian island of Newfoundland, as a result of random breeding between varieties of working dogs the early settlers brought across the Atlantic Ocean. The dogs had short-haired, water repellent coats, were eager-to-please, and worked with fishermen to retrieve any fish that fell off their hooks, as well as assisting in hauling up fishing nets (Barmore, 2008). In the nineteenth century, aristocratic English gentlemen took some of these dogs back to their home country and began breeding them selectively.

As with all the dog breeds around today, the qualities Labrador Retrievers possess are partly a product of these kinds of breeding programs. Specific animals have been chosen as possessing certain desirable traits – such as the Lab's amiability and eagerness to learn – and bred them with other animals possessing equally desirable traits in order to produce new animals that express these traits. In the case of my dog, his breeder specifically breeds Labradors who rarely bark and that are slightly shorter than the norm, since she claims that the breed was originally intended to be the same height as tall grass so the dogs could sneak around waiting to recover ducks that had been shot down. The kind of selection involved in this kind of breeding is literal: breeders really do choose the dogs with the desirable traits, then breed them together so that the resulting litter of puppies will display, to some degree, the desired behaviors and physical properties.

The other kind of selection that lead to my dog being the way he is today is the natural selection that occurred over the centuries and millennia long before his wolf ancestors were domesticated, and which give him so many of his instincts – including his powerful drive to chase and his taste for meat. This kind of selection, unlike the selection that breeders of Labradors over the last few centuries conducted, is entirely metaphorical. Selection implies choice, but in natural selection nothing literally selects which animals live or die. Rather, the circumstances of nature resulted in some animals living and giving birth to offspring, while others died and did not.

The reason Darwin originally proposed the term 'natural selection' was in order to make use of the literal selection that animal breeders undertake as a means of explaining his theory. Darwin bred pigeons, and expressly used these as an example in *On the Origin of Species by Means of Natural Selection* (1859). Pigeon fancying was a popular hobby in Victorian England, and breeders competed to produce breeds with interesting color or shapes of beak. Darwin bred pigeons in order to see how much variation was possible for a single animal, suspecting that animal breeding was in effect a sped-up version of how new species arose in nature (hence the title of his famous book). The natural selection metaphor that begins with Darwin is one of the most successful scientific metaphors ever devised – as Nicholas Thompson (2000) remarks, the grip of this image on our imagination is so strong that we now refer to animal breeding as *artificial* selection.

Selection is one of two central metaphors in contemporary evolutionary theories, the other being *fitness*. Both terms have been with the field more or less since its inception with Darwin. It is almost impossible to have a discussion on the subject of evolution without using these words, yet 'selection' was only ever a metaphor, and in so much as the 'fitness' implied by Darwin can be rendered meaningful, it must be understood as a useful fiction. Darwin's theory of natural selection is at its heart a metaphor, but as Thompson points out recognizing this "implies no disrespect for it as a scientific theory" since "metaphors play an indispensable role in science". There is a considerable risk of being misled by the imagery these terms conjure to mind, and yet it seems nearly impossible to excise them from the evolutionary lexicon.

Darwin was acutely aware of the fact that his use of 'selection' in *Origin of Species* was a metaphor. He received letters from Alfred Russel Wallace, the naturalist who was postulating a very similar theory at the same time as Darwin, expressing concern that the term 'natural selection' was too anthropomorphic, leading to a personification of nature as "selecting" or "preferring". Philosopher Michael Ruse (2003b) notes in this regard:

In his heart, Darwin seems never to have wavered, and he responded to those who criticized the term "selection" by pointing out that it was a metaphor, and who can do science without being metaphorical? "No one objects to chemists speaking of 'elective affinity,' and certainly an acid has no more choice in combining with a base, than the conditions of life have in determining whether or not a new form be selected or preserved" (Ruse 2003b, Darwin, 1868).

John F. Haught suggests that Darwin might have been more flexible in this regard, noting that the later Darwin writings sometimes seems to offer "natural preservation" as more suitable than "natural selection" (Haught, 2010), but whatever Darwin's feelings the term *selection* has certainly stuck. Darwin publicly dismissed any problems with the metaphorical aspect of the term on the grounds that science was effectively impossible without metaphorical thinking, but this does not mean Darwin was blind to the kind of constraints on thought that specific images convey. He avoided using the word 'evolution' precisely because he didn't want to take upon the baggage the term had already acquired in terms of conveying a sense of progress and destiny, an issue we'll examine shortly. Darwin seems to have believed, rightly or wrongly, that the term 'selection' could sidestep this kind of implication.

In order to fully understand contemporary evolutionary theories, it is necessary to separate – in so much as this is possible – the metaphors from the facts, the myths from the models. A conscientious audit of metaphorical terms like 'selection' and 'fitness' has much to show us about both evolution and about science in general, but this kind of critique is usually avoided, perhaps for fear of adding fuel to the fire being tended by opponents to evolution. This concern is not warranted. People are perfectly entitled to reject a particular scientific theory for whatever reason they choose, and they are especially free to object to those theories that they suspect have been ideologically contaminated. Frankly, there is little doubt that the presentation of evolution in public *has* been distorted in this way, and this by people on both sides of the fence. A defense of evolutionary studies should rest on an honest understanding of the issues, and this necessitates an acceptance of the role of imagination in its operation.

When I call 'selection' imaginary, or suggest that 'fitness' is a fiction, I *do not* mean that all evolutionary theories are mere figments, but rather that *these terms cannot be understood without reference to imagination*. Metaphor is an imaginative activity – the process of thinking about one thing by comparison to another. Science, as Darwin recognized, thrives on this kind of analogical thinking, because science – in common with the arts – is fundamentally an imaginative activity. True, much of a research scientist's time is absorbed in experiments, observation and data, none of which is enormously creative. But the experiments being conducted, and even more so the concepts that motivate those experiments, all began life as imaginative fictions. Every theory inevitably implies a story.

The fiction in which the term selection gets its meaning is that where it is *as if* something has made a decision that selects some animals and not others to survive (for natural selection) or to reproduce (for sexual selection). Because selection is intended as a scientific metaphor, it is generally considered poor form to indulge in speculations as to the obvious consequences of the fiction e.g. if we say that Mother Nature does the selecting, we've brought in a mythic figure ('Mother Nature') into what was supposed to be a sober, scientific term. This was precisely Wallace's objection to Darwin's use of 'selection' – these kinds of extrapolations follow all too easily. Darwin's counter was that it is *useful* to think in terms of selection, the story does some valuable work for us in terms of focusing our attention onto what is happening.

This brings us to the other important part of Darwin's fictional representation of how creatures change over time: *fitness*. Darwin didn't actually use the term, but did make reference to individual animals being "fitter" or being more or less "fit" than others; it is from this informal discussion of a comparative scale of "fitter" animals that the modern concept of fitness develops. Philosopher André Ariew and geneticist Richard Lewontin (2004) are very clear on the role of metaphor in this part of Darwin's ideas:

> Different individual members of a species, then, 'fit' into the environment to different degrees as a consequence of their variant natural properties, and those that made the best 'fit' would survive and reproduce their kind better than those whose 'fit' was poorer. The word 'fit' ('fittest', 'fitness') is a metaphorical extension of its everyday English meaning as the degree to which an object (the organism) matches a pattern that is pre-existent and independently determined (the environment). This metaphorical lock-and-key fitting of the organism into the environment is reflected in the modern concept in ecology of the environmental or ecological 'niche' that species are said to 'occupy'.

We can see here the fiction that Darwin was using with his original conception in *Origin of Species*: some animals 'fit' better into their environment, and these are 'selected' to survive. It is as if the world is a partially completed jigsaw, with a certain number of gaps for extra pieces. Those pieces that fit into the jigsaw persist, while those that don't fit are discarded. The heuristic value of thinking this way is comparatively clear, and the benefits can be conveyed irrespective of whatever imaginative gloss we add to the core metaphor – the jigsaw image I just proposed, for instance, or Ariew and Lewontin's lock-and-key image.

However, in the decades since Darwin, the term fitness has expanded to take onto its metaphorical shoulders far more than Darwin ever intended. In the words of philosopher of biology, Elliott Sober, fitness has "two faces" (Sober, 2000a). It not only describes the relationship between an animal and its environment, as described above, but leads a double life as a mathematical term used in formulating predictions. In Sober's words: "Fitness is both an ecological descriptor and a mathematical predictor." Trouble is, this double life threatens to wear fitness rather thin; the strain of pretending that the mathematical face is still a fit to Darwin's original use might just be too much for it to bear.

The "confusion about fitness", as Ariew and Lewontin call it, comes from ignoring a rather obvious problem. Contemporary science is in love with its numbers – empirically-minded people adore the precision of physics – but this infatuation with mathematics isn't always appropriate. Indeed, Midgley (2003) observes that our use of the terms 'hard science' and 'soft science' embed a kind of hierarchical judgment – the more mathematical a science appears to be, the 'harder' it is assumed to be, and 'hard' beats 'soft' much as rock-beats-scissors. However, Darwin's idea of *fitness to environment* doesn't lend itself to a simple numeric measure – it's a complicated, multi-faceted fiction, one that enables us to think differently about the nature of life, but only 'softly'. For most kinds of life on our planet, there is simply no way to make fitness into a neat number, much less one with predictive value – there may be no way at all to talk about one animal being 'more fit' than another outside of the fiction the metaphors provide.

Attempts to provide fitness with a robust, mathematical meaning tend to fall prey of an assumed equivalence between reproductive rates and Darwin's fitness to environment, as if 'fitness' was nothing more than a measure of the number of offspring a creature can produce. Ariew and Lewontin savage the assumptions behind this approach, demonstrating that it just won't do to treat 'reproductive fitness' as a substitute for fitness to environment, and this for a number of reasons, including the fact that a great many organisms have overlapping generations, meaning that reproductive rates are not that simple to calculate. When problems such as these have been recognized, alternative solutions have been offered – but the only secure way of mounting the concept involves measuring actual changes in the abundance of particular species, at which point any kind of explanatory or predictive power has been abandoned.

Furthermore, the attempt to use rates of reproduction as a measure of fitness run into insurmountable problems concerning what it means to count individuals at all. If the idea is that an animal which has a thousand offspring is fitter than an animal that has ten offspring (and perhaps even that the former is a hundred times fitter!) we have to be aware that there are many forms of life for which this kind of simplified view doesn't really make sense. Ariew and Lewontin offer a number of examples to clarify the nature of this problem. For instance, many flowers have offspring by seeds but also grow vegetatively by putting out underground runners that produce new flowers asexually. These new flowers are essentially clones of the original plant – should they be counted as new individuals, or not? If they *are* individuals, what does this mean for trees, which consist of a vast knot of flowering stems woven together into branches and a trunk? If the flowers are all individuals, the tree can't be counted as one single individual but must instead be treated as many different individuals, despite our strong intuitions to the contrary.

The same problem occurs with colonial organisms, such as the coral polyps that make spectacular coastal reefs. Coral polyps sometimes reproduce sexually, but more commonly they reproduce asexually via a process known as 'budding'. If fitness is to be measured by counts of offspring, is a polyp that produces a hundred sexually-produced polyps to be considered fitter than a polyp that manages to occupy an entire reef through budding? As Ariew and Lewontin put it, "the problem of fitness and relative evolutionary success demands a solution to the problem of defining an individual." They trace these issues back to the influence of Thomas Malthus on evolutionary theory (an issue we'll look at shortly) and conclude that the genetic notion of 'reproductive fitness' just doesn't match Darwin's fitness to environment.

Ariew has pursued the same issue of what fitness is supposed to mean with the philosopher Zachary Ernst (2009), and identified severe problems with every attempt that has been made to mount fitness onto a secure mathematical footing. Ariew and Ernst are all in favor of "reconstructing" the concept of fitness "so that it can play its traditional role in evolutionary explanation." They just do not believe anyone has actually solved this problem, nor indeed are they convinced that it *can* be solved. The hope is to secure a meaning for 'fitness' that will serve as an explanation for why a particular trait succeeds while others fail, and hence why animals with that trait prosper, but they are doubtful that this is possible.

Particularly at task is a widespread account of fitness that can be termed *fitness as propensity* or the propensity interpretation. The basic idea is much as we have already seen in reproductive fitness – the fitness of a particular animal is related to its expected number of offspring. To avoid all the pitfalls of this simplistic approach, however, a more advanced solution is needed which takes into account a 'family' of propensities, rather than just reproductive rates. Either way, the fitness of any given animal is taken to be a property of the animal itself, that is, of its natural propensities (hence, 'fitness as propensity').

For instance, John Beatty and Susan Finsen (1998) offer a solution to the fitness problem in terms of a family of propensities that jointly affect population growth. If all the relevant propensities are included correctly in the model, this should presumably allow fitness as propensity to function as intended – relating fitness to number of offspring by correctly modeling population growth as a consequence of multiple factors. The trouble is, under this approach, *different* propensities will be needed to consider fitness in *any* given instance, which means in order to meaningfully talk of 'fitness' we must already understand all the relevant factors that contribute to the success of particular traits.

Ariew and Ernst contend that "nature is too variegated. Different biological situations call for different algorithms to explain changes in trait frequencies." They do not see much hope of salvaging a propensity interpretation that has any predictive value. Part of the problem is that the intrinsic properties of individual animals aren't really enough to understand why particular animals thrive and others fail. For most creatures, the reasons why any given trait becomes more or less common just aren't causally determined. In the absence of a general model, it is still possible to pursue a *historical* investigation into the circumstances that lead to a particular species, but any hope that all such histories might be collapsible into explanatory theories of fitness must be set aside as untenable.

In fact, 'fitness' is applied most successfully in evolutionary theories when Darwin's fitness to environment is essentially *ignored*. Genetic models of evolution that use fitness as a mathematical term (denoted by the letter 'w') have prospered, treating fitness purely in terms of changes in gene ratios. These population genetic approaches to fitness work very well in terms of their formulation and application to specific problems in the laboratory. They can be made to fit artificial selection experiments quite easily, since 'selection' in the lab has the literal meaning of a decision (one made by the experimental protocol). But at this point, we're no longer explaining the nature of life in the same terms that Darwin was using.

What could be called *genetic fitness* in these kinds of theories doesn't assess how well a trait or a creature fits their environment at all, and as such these theories don't provide any explanation for the changes in the relative abundances of real animals or their genes. This presents a rather significant problem that tends to be ignored: if the successful use of 'selection' and 'fitness' in population genetic models has no role for the *causes* of selection, we seem to have inadvertently thrown away Darwin's theory in its entirety. These genetic models are acausal – they don't deal in causation at all, whereas Darwin's natural selection was presented *precisely* as an explanatory cause. As philosophers of biology Alexander Rosenberg and Frederic Bouchard (2008) put it: "In jettisoning fitness, acausal approaches seem to have jettisoned natural selection altogether."

We are left with Sober's 'two faces' of fitness: a mathematical model, gainfully deployed by population geneticists, and the imaginative concept of fitness to environment that originates in Darwin's theory. But these are not two faces of the same beast – they are entirely different animals, the later dealing with causes, and the former modeling probabilistic reproduction rates with no reference to causes. As Sober (2000a) notes, "fitness began its career in biology long before evolutionary theory was mathematized" – in the interim, the term has not maintained a constant meaning at all.

Of course, as Sober later notes with biologist David Sloan Wilson (2011), while mathematics has proved important in evolutionary studies "it isn't true that *only* mathematics is important, nor is it true that mathematics is *always* important." The imaginative aspect of selection and fitness may have little (or even nothing) to do with successful population genetic models, but it is still key to what Darwin's theory is claimed to say.

If, as nearly everyone involved in studying evolution would attest, differential survivorship in response to diverse environments is an important part of the history of life, then we admit that Darwin's imaginative imagery of selection, and fictional account of fitness are still important, despite having vanished entirely from the hard algebraic accounts of population genetics. Despite the supposed claims of "hard science" over "soft science", the mythology in this case somehow offers *more* than the mathematics.

The Ladder of Progress

When we think of dogs such as my Lab, or birds such as the swifts over my back yard, there is a tendency to believe that we humans are a more advanced species. Certainly we would say this of our technology, since the animals have little but their bodies and innate skills they can use while we have a vast array of tools at our disposal. Even if we manage to resist the temptation of saying that humanity is a more advanced species than dogs or birds, we might find this harder in the case of a slug, or a coral polyp, or an amoeba, or a bacterial cell. Sooner or later, the temptation to say that we are a more advanced creature becomes almost irresistible. This is the first of the myths of evolution in its most subtle form, the metaphor of the *ladder of progress* encountered as the claim that one animal is more or less advanced than another.

This problem is embedded right into the very *word* 'evolution'. As Midgley (2003) has noted, the inherent evaluative aspect of this term was the reason Charles Darwin worked so hard to avoid using it in his work, preferring instead to talk of 'descent with modification'. Despite this restraint, the connection between 'evolution' and 'progress' is almost inescapable – and indeed, marketing departments delight in using 'evolved' and 'evolution' as jargon implying major advances. The word 'evolution' always had this implicit sense of progression, long before it was applied to the history of life.

Pop culture abounds with contemporary examples of the ladder of progress. Science fiction constantly makes use of 'evolution' as a short hand for the relative degrees of progress entailed by a certain living thing. In Gene Roddenberry's movie, *Star Trek: The Motion Picture* (Wise, 1979), the entire story builds to a climax involving humanity evolving to a new, more advanced life form, a theme that reappears in *Star Trek: The Next Generation* and *Stargate SG-1* over the following decades. From the original *X-Men* comics (Lee and Kirby, 1969) through to the recent films descended from them (Singer, 2000), the theme of super-powered mutants as 'the next phase of human evolution' is quintessential to the stories being told.

We are all familiar with the iconography of the 'march of progress' epitomized by the image of a sequence of apes evolving into (for some reason) a white business man, but this connection between evolution and progress is essentially erroneous, as paleontologist Stephen Jay Gould has criticized (1995). This mythic story of progress is based upon an error inherited from Darwin's contemporary, Herbert Spencer, who used the word 'evolution' to support his belief in universal progress, which went wildly beyond Popper's milestone. This was by no means the only contribution Spencer made to the mythology of evolution. Indeed, Midgley has suggested that a great many of the viewpoints that are falsely attributed to Darwin have their roots in Spencer's metaphysics, a point we will return to shortly.

Gould states that he knows "no other subject so distorted by canonical icons: the image we see reflects social preferences and psychological hopes," rather than data or theory. He recognizes that fictional images and metaphors in science are "central to our thought, not peripheral frills" and suggests that this makes questions of "alternative representation" a fundamental element of the history of changing scientific ideas, much as Kuhn had suggested. It is this issue of alternative representations that we need to address in considering the myths of evolution – because even if we believe that the facts don't change, the way we present those facts *does* change, and different metaphors guide thinking in radically different ways.

In the context of the ladder of progress, Gould suggests that drawing a putative line of ancestors tells the wrong story about evolution, claiming that the two principal themes of natural history are diversification and stability, and these are "entirely suppressed" by this particular metaphor. We are drawn into thinking this way by considering the "the tiny, parochial pathway leading to humans", and then using this "as a surrogate for the entire history of life." Gould's assertions as to which are the key themes of the history of life is a matter of some debate, since there are many biologists who would single out different elements as worth attention, but his criticisms of the ladder of progress are a classic example of the role of scientific metaphor in theories of evolution.

In his analysis of the way metaphorical images constrain thoughts about the nature of life, Gould examines sequences of paintings intended to show the progress of life on Earth. Each picture purports to depict a different 'stage' in the development of life, but in each there is a subtle distortion of the facts. Invertebrates such as slugs, crabs and insects are shown in the early images, but as soon as the fishes arrive on the scene they become entirely absent from view – as if all the creepy crawlies had developed into fish. Yet ninety eight percent of the species on our planet today are invertebrates – they didn't go away just because creatures with spines appeared.

The theme of progress also appears in another icon used to express ideas about evolution: the cone of diversity. This is a less well-known image, but it occurs often in textbooks on evolution which show the 'tree of life' proceeding from a single trunk (the common ancestor) and then expanding into more diverse forms as time progresses. Gould notes that while the horizontal axis represents the variation in forms (diversity), the vertical axis of this model is supposed to represent time – but instead tends to be used to represent some notion of anatomical progress. The metaphor behind the cone of diversity presumes an increasing range of different species descending from a smaller set of ancestral species, but this image is thoroughly misleading.

In Ernst Haeckel's 'Pedigree of Man' diagram, for instance, humanity is situated at the pinnacle of the tree of life, surrounded by diverse mammal groups, with more 'primitive' organisms lower in the branches. Gould observes that Haeckel has fallen into the trap of believing that the vertical axis can show progress, while the horizontal spread shows diversity – thus fitting the implied iconography of a cone of diversity, Haeckel takes mammals – a small group of some four thousand species – and makes fine distinctions into whales, carnivores, primates etc., while insects – representing almost a million different species – are forced to occupy a single twig since this more 'primitive' form of life must be fitted into a lower level.

Gould suggests that rather than a tree of life, the fossil record rather suggests a bushier 'plant' imagery. His basis for this approach lies in a period of prehistory known as the *Cambrian explosion*, a comparatively short interval (at least as far as geological time is concerned!) about 500 million years ago during which multi-cellular life appeared in myriad forms. The weird and wonderful fossils from this period, which Gould spent some time studying, show a wealth of different body plans, all of which seem to have appeared with astonishing rapidity as far as the usual time scales for evolution are concerned. Paleontologists are by no means united in their interpretations of the diversity from this period, but they are in agreement that something significant did indeed occur at this point in time – the metaphor of an 'explosion' of diversity is not generally challenged.

Gould offers an alternative diagram showing a massive spread of diversity a short distance up the trunk (representing the Cambrian explosion) with just a couple of the many stems depicted in his image continuing on to the present, splitting into small branches on the way. He referred to this as an "icon of a grass field with most stems mowed and just a few flowering profusely". This reflects a central claim in Gould's accounts of the history of life, namely that a great many of the species that have ever lived on our planet are now gone forever, as a result of a kind of *extinction by lottery* that in no way suggests any sense of destiny or progress.

This relates to the evolutionary theory most associated with Gould, the one he developed with Niles Eldredge (1972) known as *punctuated equilibrium*. According to this model, rather than evolution occurring more or less constantly, the bulk of the history of life on this planet can be characterized by long stretches of stasis, in which nothing very much changes at all. Periodically, there are mass extinctions or other great upheavals which 'punctuate' the otherwise static circumstances of the natural world. By killing off vast numbers of different species, effectively at random, disasters like asteroid impacts create opportunities for new kinds of life to appear.

Gould's 'mowed grass' or 'bush' metaphor stresses the role of contingency – luck, if you will – in the development of life. "All our canonical icons are based upon the opposite notion of progress and predictability," he accuses (Gould, 1995). Of course, this alternative mythology of evolution embeds Gould's own preferred metaphors – the emphasis on diversity (the Cambrian explosion) and stability (punctuated equilibrium) in particular. According to Gould, the reason we see progress when we look at the past is that we naturally interpret the natural history of life in terms of the story of how we got here – even though objectively, there is no way we can validate any kind of claim to progression in nature.

However, there is nothing close to a consensus of support for this alternative viewpoint on the role of progress. Indeed, Edward O. Wilson (1992), one of the most prominent figures in discussion of evolution in recent decades, was insistent that there was indeed some kind of advancement to be detected in the history of life:

The overall average across the history of life has moved from the simple and few to the more complex and numerous. During the past billion years, animals as a whole evolved upward in body size, feeding and defensive techniques, brain and behavioral complexity, social organization, and precision of environmental control – in each case farther from the nonliving state than their simpler antecedents did... Progress, then, is a property of the evolution of life as a whole by almost any conceivable intuitive standard, including the acquisition of goals and intentions in the behavior of animals.

Part of Gould's argument against progress rests on the claim that if we could rewind the clock and let evolution take place again, we could not hope to expect the same results. This is why his metaphors embed a critical role for chance (extinction by lottery) – rather than evolution necessarily leading towards more advanced forms of life, Gould insisted that we simply could not know what would happen if we could repeat the 'experiment'. But there is a limit to this kind of claim, since whichever evolutionary theories we happen to believe, we are in no position to know with any certainty what might result from a rerun of life. By necessity, this deals with untestable situations, and whether we believe Gould's claim that it life-rerun would be different or not, we cannot know with any certainty. The answer lies far beyond Popper's milestone.

However, perhaps we should not be so hasty in reaching a conclusion. While the outcome of a complete rematch of the game of life is certainly not on the cards, we can find some evidence in the world today that might be used to argue against Gould's claim that an entirely different outcome is the only plausible result. Fryer and Iles (1972) famously studied the many different species of fish in the great lakes of Africa, particularly Malawi and Tanganyika, and discovered that despite each environment being entirely separate and distinct, the fish in each lake were remarkably similar. There seems to have been massive convergence, with many different species developing in highly similar ways in each separate lake – in direct contrast to the theme of contingency that is central to Gould's claims.

Furthermore, paleontologist Simon Conway Morris (2003) has argued that the kind of patterns found in the lakes of Africa reflects a wider tendency in the evolution of life. Different lines of descent tend to converge in similar adaptive solutions, and as such, even humanity might not be a freakishly unlikely occurrence. (Again, however, we can't really know, since whatever the evidence in other species, speculation about our own evolution is deeply metaphysical). Conway Morris expressly argues against Gould's interpretation of the Burgess shale, the seam of rocks in which fossils from the Cambrian explosion are most abundant, and suggests that the number of "end points" for evolution has severe limits such that "by no means everything is possible". It is his view, rather controversially, that "the emergence of the various biological properties is effectively inevitable."

Theologian John F. Haught (2010) suggests, in a kind of contemporary spin on the ladder of progress, that whatever standard of measurement we want to use "it is hard to deny that our universe has come a long way from the relative simplicity of primordial cosmic plasma to the nearly incalculable complexity of the human brain and cultural innovation." He suggests it would be "highly arbitrary" to ignore what seems to be a radical directionality. Haught has his own religious motives for finding a pattern in the history of life, but it is striking that Wilson too felt an undeniable progress was detectible in nature, since his beliefs are radically distant from Haught's. The recognition of progress in evolution seems to be widespread.

Ruse (2003b) also suggests that despite the random factors that affect evolution, the trend is consistently slanted towards greater complexity, and he offers a metaphorical image as an explanation for why, despite contingency, there still seems to be a discernible pattern of progression. He suggests the evolution of life may be like a "drunkard on a sidewalk bounded by a wall on one side and the gutter on the other." No matter how inebriated he might be, the drunkard cannot go through the wall – but he will eventually end up in the gutter. "Perhaps," Ruse suggests, this is what we can see in the history of life – "a random stagger toward complexity."

On the whole, as Ruse admits, the jury is still out on the ladder of progress. On the one hand, there are many who side with Gould in believing that contingency is so critical to the development of life that it is utterly misleading to read progress into its history. We only see ourselves as the culmination of evolution because of our arrogance and self-centeredness. On the other, however, biologists, philosophers and theologians with few other beliefs in common seem willing to defend some version of the ladder of progress myth – whether in terms of convergence, a 'drunkard's stagger' or just plain intuitive hunches. There is no consensus on this theme, and since the big questions rest on entirely untestable metaphysics, we perhaps cannot hope to settle this issue at all.

Perhaps the solution is not to debate the validity of progress as a means of understanding the nature of life, but to consider a different approach entirely. Even if legitimate claims can be made to some kind of advancement having occurred throughout the history of life, the ladder of progress still tricks people into imagining the process can be projected forward, when in fact it can only be selectively represented from the past. Science fiction stories of 'the next step in human evolution' show just how misleading the ladder iconography can be, and this alone argues for a different spin, one not so prone to radical misunderstandings.

We can gainfully consider our evolutionary heritage not in terms of a metaphor looking from the past to the present, as the ladder does, but from the *present to the past*. By looking back, we remove the temptation to illicitly project forward, and gazing into the past the best image isn't a ladder but rather a *chain of inheritance*, an anchor to the past from which we gain all our biological gifts. What comes in the future depends upon the conditions that will come – and this can never be predicted – but embedded within all life are aspects of the species that came before, a sequence of connectivity that goes back even to the single-celled organisms that for two thirds of the history of our planet were the only life to be found on our planet.

How-Why Games

Why are polar bears white? Although it is easy to offer logical explanations to this question using the metaphors of natural selection – such as 'white fur provides a selective advantage when hunting seals' – to do so is to play a special kind of game, one that we should be cautious of accepting too readily. Such stories invoke what is called *teleology*, which is to say, an explanation in terms of purpose or design. Before Darwin, this kind of *teleological* or design-focused thinking was the dominant explanatory mechanism in biology, explaining features in terms of why it was the best solution. Since Darwin, the same kind of thinking has *remained* the dominant explanatory mechanism – all that has changed are the justifications that are attached to these explanations.

What is wrong with saying 'the polar bear's fur is white because it confers a selective advantage when they are hunting seals'? The answer to this is simple: how would we test this claim? In so much as an assertion like this can be taken as a scientific hypothesis, it must be viable to provide some kind of evidence or experiment that could potentially disconfirm this explanation. If this isn't the case, if there is no way to test it, there is no way to falsify it and (following Popper's critique) we have to admit we are dealing with metaphysics and not empirical science.

The vast majority of casually stated claims of the kind above are what I term *How-Why games* (or *teleological games*). The name relates to a particular kind of traditional story that is told in the form "How the Elephant Got It's Trunk" or "How the Rabbit Got

Its Tail" – folktales that are presented as an explanation for *how* some feature of the world came to be the way it is. In so much as these claims are mounted in terms of an interpretation of design, or an imputation of purpose (that is, in terms of teleology), they are not just stories of how something came to be, but also purportedly scientific statements concerning *why* a certain aspect of biology is the way it is. This is why I call them How-Why games – they combine the "how" style of a folktale with the "why" style of scientific explanation.

Let's consider the How-Why game of the polar bear's white fur to see just why this kind of explanation can be understood as folktales dressed up as science. Consider, for instance, the scenario that at some point in the past the polar bear was hunted by a predator. In this story, we can claim the polar bear is white as a consequence of a selective advantage that helped them avoid predators. But then, consider the scenario that polar bear ancestors became white purely by chance, and then discovered that being white allowed them to hunt seals. The hunting of seals has become an effect when previously it was suggested as a cause!

Science does not advance by fabricating logical explanations and then accepting them because they sound reasonable – empirical studies expressly presume that when we make a statement we can test and (to some extent) verify that claim experimentally. Claims that cannot be tested lie beyond Popper's milestone no matter how logical they sound. As Nicholas Thompson (2000) charges, if we do not anchor our explanations for biological features in appropriate systematic methods "a theory can easily degenerate into a series of ad hoc explanations." We need to be particularly careful about using evolutionary theories as a basis for constructing fanciful-yet-plausible folktales, since it is all too easy to be drawn into utterly absurd beliefs about how life came to be the way we observe it.

Anyone can play a How-Why game – and it's fun to do so!

Simply look at some animal and devise an explanation for its features or behavior in terms of some explanatory scenario. Those players that do not wish to use evolution as the source of the resulting folktale are free to substitute God as the explanatory factor, while those that would rather play *sans Deus* might prefer to use 'selective advantages' or something similar. The game will play quite similarly either way. Consider this example from a book on seashells written by Mary Roberts (1834) in the early nineteenth century:

An ovate or oblong form is consequently the very best that could be adopted; and, moreover, the points with which it is covered and adorned, are evidently designed to protect the shell from external injury... At the same time a beautiful variety of tints evince that minute attention to the finishing and decorating of his works which the Deity so continually displays.

Roberts plays her How-Why game with God, claiming that animals display the best form that could be adopted because of the care and attention of God. (We will examine these kinds of design explanations in greater detail in a later chapter). Contrast this with a letter from Joanna Burgess to *New Scientist* from the late twentieth century (Burgess, 1999):

Many citrus trees that are natives of arid regions have sour fruit to discourage animals from eating it. The flesh of a lemon is there for three main reasons: to add weight so it will roll a long way after it falls from the tree, to dissuade foraging animals from eating the seeds before they can develop, and to supply water and nutrients as the flesh rots around the germinating seeds. The main aim of any seed is to propagate the species, not to feed the local animals.

Burgess plays her How-Why game with the *gene-centered view* (which we will explore more thoroughly in the next chapter). She devises an explanation for sour fruit on the assumption that "the main aim of any seed is to propagate the species". But this is an odd claim! Because what does it mean to suggest that a seed has 'a main aim'? Surely we aren't supposed to believe that the fruit or the fruit tree actually possess motives or minds – this is the sort of fantasy we expect in children's cartoons, not in scientific discourse. Burgess proceeds from the assumption that the meaning or purpose of life is to propagate itself, but this is surely metaphysics, for any discussion of 'meaning' or 'purpose' must necessarily fall outside of empirical science.

It is certainly logical to suggest that fruits in arid regions became sour because it was a selective advantage to discourage animals from eating them, thus keeping their precious water to themselves. Elsewhere, where water is more common, having tasty fruit that animals enjoy eating can help spread seeds over a wide area, thus also offering a selective advantage. But does this kind of story really make sense? Are we suggesting that once upon a time all the fruits were tasty, but the desert fruits adapted to being sour? What if once upon a time all the fruits were sour, but the temperate fruits adapted to be tasty to gain the opposite advantage? There are so many possible stories we can tell, *and few if any of them are testable.*

The problem occurs in part because of the assumption that evolution is purely a game of competition: species compete with each other for resources and the 'fittest' (i.e. the best adapted to its environment) survives. But even if this assumption were true, the lineage of any given species spreads over millions (even billions) of years and we have to rely upon the utterly incomplete fossil record and the equally sketchy DNA evidence to provide clues as to what was a selective advantage at any given time. We cannot know which traits provided specific selective advantages, because we cannot see through time, and nothing in our evolutionary theories allows for us to predict that *all* observable traits provided an advantage at all times, or indeed at any given time.

Gould's metaphor of extinction by lottery served to represent his view that chance played as big a role in the history of life as competition – that evolution was not just a sporting contest, but also something of a gambling game. This argues against using competitive arguments as a necessary explanatory framework, since we can't distinguish between outcomes dominated by luck and those dominated by 'skill'. Furthermore, as Richard Lewontin (1991) notes, the usual process in biology of looking for *the* cause of an effect – presuming that there is such a thing as a major cause, and that all other factors can effectively be ignored – is a terribly naïve way of looking at biological systems which contain many intricately interrelated elements, and do not lend themselves to analysis through overly simplistic models of causation.

As I mentioned before, Popper suggested that Darwinism was a metaphysical research program and not a testable scientific theory. He later retracted this claim because he was shown evidence of biologists using optimization analysis to make predictions about changes in the statistical distributions of characteristics in real animal populations. But while the capacity to make testable assertions about *future* populations constitutes a validation of Darwin's theory, it categorically does not mean that How-Why games played using Darwin's theory (or something similar) are any more scientifically valid than those same games played with God. The validation of the theoretical framework is an entirely separate issue from the widespread practice of using metaphysical beliefs to invent folktales and then present them as if they were empirical science.

We have several sources of data from which to form scientific theories about the past. The fossil record provides evidence spanning billions of years, but it is incomplete, and its interpretation can be highly subjective (as clearly indicated by the case of the Burgess Shale, which Gould and Conway Morris interpreted differently). The geological record is more complete, but also more general. The information we can glean from examining genetics via molecular biology provides some additional information, especially about events less than a million years ago (such as historical human migrations), but it does not allow us to see very far. Radiocarbon dating is another source of data, but only about the ages of things. It is likely we will acquire some new techniques over time, but we will always be interpreting the past with a mere sliver of data, at least compared with the wealth of options we have for the scientific investigation of other subjects.

We have to be careful when we interpret our observations of the natural world, because the lure of a How-Why game is always there. As Richard Dawkins (1995) has noted, human beings may have "purpose on the brain" – we tend to see patterns and explanations, even in situations that are hard to justify. Midgley (1992) remarks that reasoning from purpose (i.e. teleology) is "a much more pervasive, much less dispensable element in human thought than has usually been noticed." She suggests that our imaginations might not work without it, since thinking in terms of purpose (and consequently, in terms of design) is "woven into all our serious attempts to understand anything", and this is especially the case in science. The dangers of letting this natural tendency overwhelm us is that we risk confusing the matter at hand since, as Lynch (2007) observes, "evolutionary biology is not a story-telling exercise" – we have to be careful to distinguish folktales from empirical science.

The kind of explanations entailed by a How-Why game may feel scientific because they seem to be invoking a kind of ordered, law-like justification for how and why things have turned out the way they have. We associate these kinds of explanations with empirical science, because so many scientific theories do indeed offer rigidly defined mechanisms. But when it comes to the history and nature of life, these kinds of accounts are not necessarily appropriate. As biologist John O. Reiss (2005) has suggested, when we turn to purpose-focused thinking we are necessarily explaining events that are *contingent* for us – because if they were determined by entirely known laws there would be *nothing left to explain*. A failure to detect this distinction causes How-Why games to be treated as empirical science, rather than a fanciful excursion out beyond Popper's milestone. The kind of explanations being offered are "at best a sketch of the possibility" of a more robust explanation, "not such an explanation itself".

In the sense of identifying a single causal factor, we may never be able to say with any scientific confidence 'why the polar bear is white', although we may conjecture to our heart's content! We are free to play our How-Why games however we wish, but we should not present them as if they were valid scientific results. When we make statements without a framework that could allow them to be tested and falsified we are blurring the boundaries between metaphysics and science and thus ignoring Popper's milestone in order to skip merrily into unknowable terrain. Ideology – religious, scientific or otherwise – has tremendous power to distort our understanding of the nature of life, and can cause us to buy into stories of biological destiny that are oddly akin to ancient folktales.

Survival of the Fittest

There is a long standing adage often associated with Darwin's theory that states that 'only the strong survive'. This is a rather odd claim, however, since it is abundantly clear that the natural history of our planet does not tell the tale of the more powerful creatures outlasting their weaker competitors. Just consider the case of the dinosaurs: tyrannosaurus rex, the poster child of bygone scary monsters, was thriving at the very end of the Cretaceous period but despite its power, it and dozens of other carnivorous therapods were destined for extinction, while its smaller, weaker relatives survived to become the birds we still see around us today.

The idea that strength is the most prevalent survival trait is not even remotely based upon scientific observations, and draws primarily from people's political and economic beliefs about competition. But even accepting the rather limited view that competition is the sole aspect worth focusing upon (mistaking ubiquity for quintessence), strength is just one of many competitive advantages that can help a species – or an individual – survive. Take the swifts: their dinosaur ancestors clearly survived the termination of the age of the terrible thunder lizards, but they did not do so because they were *strong*. In fact, we have no way of knowing just *why* they did survive, although we can play How-Why games to our heart's content.

The mythology behind this idea that 'only the strong survive' is directly connected with the second myth of evolution: *survival of the fittest*. So engrained is the imagery of this phrase that it is frequently used as a means of connecting an arbitrary competitive situation to evolutionary theory – but this connection is almost always spurious. 'Fitness', as we have already seen, has a number of different meanings, but none of them will validate the idea of survival of the fittest as a useful description of nature. In Darwin's fitness to environment, for instance, there is no 'fittest' to speak of, since the 'fit' he is talking about is the correspondence of an animal to its environment. One creature can be a better fit than another – it can be 'fitter' in any given context – but that doesn't mean that we can single out animals that are 'fittest'. The other legitimate use of 'fitness' is in the context of genetic fitness – but here the phrase 'survival of the fittest' can't do much useful work since the molecular biologist's term 'w' is specified as a measure of the differential survivorship of genes, so to say that genes with high w values survive is simply to restate its definition.

The misleading stance on the nature of life embedded in the

myth of survival of the fittest does not, in fact, originate with Darwin at all. The phrase was coined by Darwin's contemporary, Herbert Spencer, and was in use in Victorian economics before *Origin of Species* was in print. Spencer's viewpoint was distorted by his prior commitment to the primacy of progress, and hence the first myth of evolution relates quite explicitly to the second, but there is a sense in which Spencer's phrase relates not so much to biology, but rather to *sociological* metaphysics. Those views that are based on Spencer's beliefs are generally called 'social Darwinism', but as Midgley (2003) notes it would be better called *Spencerism*, since Darwin himself rejected this approach entirely.

Spencer was using the term 'evolution' before the publication of Darwin's landmark book, and believed he had found a "Law of Evolution" that could be used to justify political freedom in the marketplace. He contended that free trade would ensure "the survival of the fittest" and dubbed the resulting principle 'evolution', a term he was largely responsible for popularizing. It was partly because of this deployment of evolution as a justification for highly competitive, unregulated economics that Darwin was very careful not to use the word 'evolution' when writing *Origin of Species*, although he does mention it once at the end. Ironically, Wallace pressured Darwin to accept Spencer's phrase on account of his concerns about 'natural selection' being too anthropomorphic and teleological, and his belief that 'survival of the fittest' was less likely to be "misrepresented and misunderstood". Indeed, later editions of the *Origin* did include Spencer's phrase (Ruse, 2003b).

It's not clear whether Spencer's 'Law' would have caught on had Spencer himself not decided that Darwin's theory supported his doctrine. There was no real connection between Darwin's theory and Spencer's ideology, but having claimed that there was, Spencer was able to spread his economic beliefs even more effectively to a wider audience. As Midgley states, it was "under the banner of Darwinian science that Spencer reached, and converted to his views, a large and receptive audience, especially in the United States." This was disastrous not only because, as recent events have confirmed, unregulated markets can become dangerously unstable, but also because the connection between Spencerist economics and Darwin's natural selection has contributed to the popular belief that evolution is fundamentally immoral.

The ideology behind Spencer's survival of the fittest myth had its roots in the ideas of the eighteenth century clergyman Thomas Malthus, who had been concerned that resources were effectively limited, and hence population was a problem that needed to be addressed. Ariew and Lewontin (2004) confirm that in some sense there was a "Malthusian origin of evolutionary theory" in their discussion of the problems of numbering individuals. But population numbers are not really an adequate measure in this regard, since success in the context of limited supply is about how much of a resource can be sequestered, not about the numbers of individuals. They observe that "numerosity is a proxy for total resource, but it is only a proxy." In the absence of a suitable means of measuring resource acquisition, it's essentially impossible to talk about fitness as a measure of success in the way that 'survival of the fittest' presupposes.

Additionally, there is nothing testable in the idea of the primacy of competition, and the Spencerist ideology has lamentably been used to fuel all manner of abhorrent ideologies such as imperialism and racial supremacy. As we saw in the previous myth of evolution, luck can have just as great a role in influencing biological evolution as strict competition, and the same is true in the social or economic realms. The people and companies that succeed certainly have advantages that they leverage well, but this doesn't mean that they were necessarily superior in any sense to those that failed – no-one is immune to a bad break. Particularly in the case of companies, the difference between astronomical success and crippling failure sometimes comes down to nothing more than fortuitous timing.

Furthermore, it is readily apparent that even recognizing numerous aspects of competition in society doesn't preclude a parallel recognition of the benefits of co-operation: if companies compete in the national marketplace, and nations compete in the international marketplace, it is important to remember that both the company and the nation represent examples of widespread co-operation – if this were not the case, all trade would be between individuals. The same is true of nature. Whereas 'survival of the fittest' may suggest Tennyson's image of "nature red in tooth and claw", Midgley (1985) notes that "nature is green long before she is red, and must be green on a very large scale indeed to provide a context for redness."

It is wise to recognize that survival in nature is not solely a matter of exerting superior force or dominating by violence. Although the most powerful (or deadliest) species in any given habitat generally will become top predator within that particular ecological context, survival for apex predators is extremely precarious. Animals in such a position depend upon the robustness of the entire food web they are embedded within for their survival. A fox may be stronger than a rabbit in terms of the capacity to cause harm, but if something threatens the survival of the rabbit, the fox is equally threatened. Predators inherit vulnerability from the species they prey upon when ecological equilibrium is lost. Rather than generating survival advantage, predators actually suffer extreme survival *disadvantages* during times of crisis, precisely because they are dependent upon the success of their prey species.

Contrary to the claim that 'only the strong survive' the trait most suited to the long-term survival of a lineage of animals it is not strength but *flexibility*. Few predator species rack up more than a few tens of millions of years at the top before becoming extinct: as the fossil record repeatedly attests, the bigger and stronger you are, the harder you fall. At the end of the reign of the dinosaurs, some sixty five million years ago, it was the tiny yet adaptable mammals that gained the edge (roughly ten of the fifteen mammal families around at the end of the Cretaceous survived), along with some of the smaller, more adaptable therapod dinosaurs that diversified into modern birds. Flexibility is the safest long term strategy, because being strong only gets you as far as the next cataclysm.

Survival of the fittest, therefore, offers a thoroughly misleading perspective on the nature of life, radically overemphasizing the role and importance of competition. This has serious knock on effects on the way that natural selection is discussed. How-Why games conducted on the assumption that natural selection necessarily works as "cut-throat competition" are pursued obsessively, without any consideration of the phenomena to be explained (Midgley, 2011). It is simply assumed that there must be hidden causes for everything that are explicable in terms of survival advantages – taste for mathematics, jokes and laughter, music and poetry, play, sympathy and the tendency to quarrel (to cite examples Midgley offers). This quest is "perverse and empty" since these are all "done for their own sake because they fulfill our nature."

The myth of survival of the fittest seems to say: survival is a game and it is won by those who are best at playing it (the 'fittest'). As a story to encapsulate what evolutionary theories teach us about natural history, it is remarkably ill chosen. Competition for resources and the struggle for survival are indeed a key part of natural history, but competition for mates is just as important as competition for survival in driving the selection process, as Darwin himself recognized in his concept of *sexual selection* (1871). In many social animals the ability to co-operate to protect the young is radically more important than intra-species competition, and other forms of co-operation even between indirectly related animals are equally significant.

Survival is not a zero sum game in which every winner implies a loser, but this fact is obscured by the 'fittest' mythology. Except in a purely trivial sense where 'fittest' means 'has survived', this myth is rather empty of applicable meaning.

If we look at the actual history of life on our planet, what is suggested is not the brutal gladiatorial conflict implied by 'survival of the fittest'. Rather, the history of life on our planet reflects a continuous yet periodic *refinement of possibilities*. Just as the breeders who selected the ancestors of my dog refined the possibilities of his breed, so nature has refined the possibilities in all creatures alive today. The intense diversity of life is also a product of this relentless inventive refining process – each species reflects different possibilities, and the number of possible kinds of creature has continued to increase over the billions of years since life first appeared.

In offering this alternative myth I do not seek to deny that competition for resources or the background struggle for survival are a major part of life in the wild, nor that natural selection seems a harsh process when viewed long-term, but focusing solely on these aspects of the history of life risks radically curtailing the scope of our understanding – seeing only the warp of the tapestry of life, and missing the weft. The fox preys upon the rabbit, but in doing so the possibilities inherent in rabbits become refined – and in reacting to these refinements, the possibilities of the fox are also gradually refined.

Life on our world has sprung into diverse existence and been savagely curtailed by disasters and extinctions, but throughout the unfathomable duration of geological history it is clear that evolution has created new possibilities, new and amazing forms of life, as well as novel ways for these animals to interact. Some see in this story a kind of inevitable progress, and some do not, but no-one who accepts evolution can deny that everything alive today traces its ancestry through radically different forms of being. Seen through this lens, our evolutionary history tells the story of the gradual refinement of the possibilities of life – from which process emerges all the many living things of our planet, including you and I, the dog I take to the park, the swifts in the skies above my garden, and everything that will follow in our wake.

3. Gene Supremacy

The Gene Confusion

One evening during the latter stages of working on this book, I went down to the kitchen to get something to drink and noticed a cat behaving strangely in the back yard. I've spent a great deal of time with domestic cats over the years, and I could see that there was something particularly strange about the way this one was acting. Ordinarily, cats look through the plate glass windows at the back of our house to see whether my dog is going to burst out and chase them away from his territory, but this one was clearly not concerned about a canine intervention on this occasion. No, it was definitely hunting something.

I went outside, and the cat backed off a little. I looked around to try and see what it was stalking. Sure enough, hiding under the wheeled recycling bins was what I at first mistook for a white mouse that had lost its tail – possibly in the fight with the cat. It was clearly a domestic pet and not a wild animal so I fetched my wife and we escorted our rodent guest into the safety of the house, where it recovered from its ordeal in an empty shoebox we'd lined with newspaper. We put up posters to see if anyone had lost a pet mouse, and discovered from the local pet shop owner that what we had found was actually a hamster – rather than losing its tail in its skirmish with the cat, it (thankfully!) only had a stub of a tail to begin with. The hamster has ended up becoming a part of our family – even the dog plays with him when he's out-and-about in his plastic hamster ball.

There are any number of ways that we can talk about the incident between the cat and the hamster that took place in my back yard. No-one would dispute that the cat, in pursuing the hamster, was only following its instincts – the desire to hunt small animals in general, and rodents in particular is something found in domestic cats around the world. Similarly, no-one would dispute that the hamster, in sheltering under the trash cans, was following its instinct to hide in narrow and dark places when threatened. However, disagreements begin the moment explanations for those instincts are discussed. According to one camp of evolutionary biologists, everything that happened between the cat and the hamster was ultimately caused not by these animals themselves, but by their genes.

In 1976, Richard Dawkins published his book *The Selfish Gene*, which built upon George C. Williams ground-breaking idea that 'adaptation' was too vague a term to build a scientific theory upon, and that evolution could be better understood as selection among genes. Dawkins later helped develop this concept into what is now called the *gene-centered view*. This perspective provides a unique view of the history of life and no-one can truly claim to have grasped modern theories of natural selection without some appreciation for how things seem when considered from what Dawkins has called "the gene's eye view" (Brockman, 1995). He recognizes that the idea is not really his, but states that he has "done the most to sell it", seeing his contribution in terms of adding rhetoric, and spelling out its implications.

Dawkins admits that he is not a geneticist, nor "particularly interested in genetics", but in trying to teach Darwin's theory in the context of animal behavior he was struck that "the most imaginative way" of looking at the subject, and "the most inspiring way of teaching it" was to suggest "it's all about the genes". It is his view that the genes, for their own benefit, "manipulate the bodies they lie about in" and that each individual animal is merely "a survival machine for its genes" – a "robot" that is programmed in advance by its genes and "carries around its own blueprint, its own instructions". The genes that survive are those that "managed to make their robots avoid getting eaten" long enough to reproduce.

According to the way Dawkins sees animals, therefore, the cat and the hamster came into conflict because each had been programmed in advance by their genes in order to behave in particular ways. Dawkins isn't trying to say that this *specific* fight between a feline and a rodent "robot" was pre-programmed, but rather that in trying to ensure their propagation into future generations the genes of the cat and the hamster had specified instincts geared towards the survival of these "machines" in order that they might breed and pass genes onto another generation. Hence the cat hunts rodents because wild cats instinctively preyed upon animals like the hamster for food, and in so doing were able to live long enough to breed and pass their genes onto another generation. Similarly, the hamster hides from the cat because wild hamsters instinctively hide in burrows in order to live long enough to breed and pass on their genes to their descendants.

This science fiction description of animals as mere robot puppets whose sole reason for existence is the propagation of their genes, is the third myth of evolution: *the selfish gene*. As is so often the case with the mythology of science, there is a legitimate empirical perspective tied up with the imaginative metaphor and the *gene's eye view* does serve an important role in contemporary understanding of evolution. However, the way this concept is presented in Dawkins' metaphor is rife with misconceptions. Two stories in particular can be implied from this imagery, firstly the myth of *egoism*, which claims that all behavior is fundamentally selfish, and secondly the myth of *gene supremacy*, an over ambitious dogma that mistakes an explanatory principle for a fundamental law.

Interestingly, Dawkins is aware that his work involves inventing imaginative imagery. Indeed, he believes that rather than popularizing scientific ideas he should be understood as "a creative force" whose ideas "influence and change people's lives – change the way other scientists think, make them think in a different, constructive way" (Brockman, 1995). In this respect, he is in agreement with critics of his work such as Mary Midgley, who also recognizes the imaginative aspect of Dawkins work, noting the "lush mythology of gene-selfishness" that Dawkins has created (Midgley, 2011). There is no doubt, therefore, that Dawkins is engaged in scientific myth-making – the question concerns the nature of those myths, the extent to which they legitimately present scientific beliefs, and how well they represent the nature of life.

It is perhaps prudent to get some perspective on the valuable side of Dawkins' mythology before pursuing its problems in more detail. Michael Ruse is typical of those who find the selfish gene to be helpful, calling it a "felicitous metaphor" (2003b). While he recognizes that "talk of selfish genes is surely pushing reductionism – explaining the smaller in terms of the larger – to the extreme" he also recognizes that genes are indeed central to evolution, to the extent that he states: "No gene change, no evolution". It is his view that Dawkins' talk of selfish genes is "rhetorically dramatic" but "hardly all that new or peculiar". After all, as Dawkins himself states, the roots of his approach lie in neo-Darwinian synthesis of the 1930s, particularly biologists such as W.D. Hamilton and the aforementioned George C. Williams. Ruse suggests that even if there is a sense in which "the modern Darwinian" sees selection as "counting the proportion of DNA copies that gets poured into the next generation", this doesn't mean that the animals themselves can be ignored.

Others take a more lukewarm position. W. Daniel Hillis speaks positively of Dawkins' imagery as being "powerful and exciting", encouraging new ways of thinking, but also complains that he spends a lot of time arguing against people who have become carried away with this kind of picture since they are "too easily misunderstood as explaining more than they do" (Brockman, 1995). Niles Eldredge similarly suggests that the selfish gene metaphor became "a kitchen industry" which

Dawkins has been "mostly responsible" for spreading (Brockman, 1995). Many other scientists are in a similar position, seeing Dawkins' contributions as helpful on the one hand, but also somewhat misleading on the other. Nicholas Thompson, for instance, suggests that the metaphorical use of 'selfish' in 'selfish gene' is vague and unhelpful. He sees Dawkins work as "an important corrective" to the idea that natural selection made altruism implausible, but recognizes that it can also be highly misleading (Thompson, 2000).

The most scathing attacks that have been directed at Dawkins from a fellow scientist came from Stephen Jay Gould, and indeed the two scientists fought bitter public battles over their differing interpretations of evolutionary mythology. Gould savaged Dawkins' obsession with genes, claiming that the gene-centric view was "a confusion of bookkeeping with causality", providing an absurdly reductionist perspective which he characterized as "Darwinian fundamentalism" (Gould, 2002). Gould's criticisms in this regard somewhat overstepped the mark, probably as a result of the anger and hostility engendered by the heated conflict between the two highly regarded scientists.

However, it should not be thought that Gould's critique was entirely without warrant. As Midgley (1985) has repeatedly suggested, Dawkins' rhetoric all too frequently involves "elevating the gene from its real position as a humble piece of goo within cells to a malign and all powerful agent." There is perhaps no more striking example of this mythic inflation of the signifi-cance of genes than the opening assertions of *The Selfish Gene* itself:

The argument of this book is that we, and all other animals, are machines created by our genes... Like successful Chicago gangsters, our genes have survived, in some cases for millions of years, in a highly competitive world. This entitles us to expect certain qualities in our genes. I shall argue that a predominant quality to be expected in a successful gene is ruthless selfishness... If you wish... to build a society in which individuals cooperate generously and unselfishly toward a common good, you can expect little help from biological nature. Let us try to teach generosity and altruism, because we are born selfish.

This is a wild piece of science fiction! The genes are cast in the roll of criminal masterminds ("Chicago gangsters") who are secretly in charge of everyone, while animals are relegated to mere pre-programmed robots. As this chapter seeks to demonstrate, it is a strange claim that because particular gene variants have survived that we can expect them to be "ruthlessly selfish", since the fact that a particular kind of gene has survived for millions of years entitles us only to conclude that it was *useful* to the creatures that possessed it. Furthermore, the idea that we are "born selfish", as we shall shortly discover, is completely outside of Popper's milestone and thus has a very weak claim to scientific validity.

One might defend Dawkins story on the grounds that is merely a piece of hyperbolic sensationalism intended to grab the readers' attention, and that is certainly the case. However, it is hard to escape the sense that Dawkins has drunk his own Kool-Aid, so to speak. He writes in a later book that while DNA "neither cares nor knows" when it comes to genes "we dance to their music" (Dawkins, 1995). Not only are animals presented as mere vehicles, genes are offered as "a kind of active hypnotist doing the driving" (Midgley, 2003) What's more, by suggesting that "genes are the immortals" (Dawkins, 1976) types are confused with entities. After all, each specific gene is lost when the cell it belongs to dies. If we can talk of genes as immortal because their *patterns* persist, we might equally talk of humanity as immortal because it is "represented successively by different individuals" whilst "transcending and outlasting them all"

(Midgley, 1985).

Despite Dawkins' suggestion that "genes have ultimate power over behavior" (Dawkins, 1976), the relationship between genetics and behavior is not even remotely that simple, and we will go quite awry if we fall under the delusion that our own behavior is beyond our control. Consciousness cannot simply be abstracted into insignificance, and motives belong to beings with minds, not to self-replicating chemicals. Genes are nothing more than DNA patterns for constructing proteins, and although these proteins are indeed used to construct behavioral systems, including elements of the brain, neurotransmitters and hormones, the gene is just a component of the template for an organism's biology. Just as a brick used to construct a building tells you little about what people do in buildings, a gene helps build a body but by itself tells you little about what that body does. Behavior – even among simple animals – depends as much on environment and culture (or ecology for less complex creatures) as it does the biological capacities inherited via genetic transmission.

A brief history of the gene may be helpful. In the 1860's Gregor Mendel experimented with cross-breeding pea plants, and came up with principles of inheritance that are still taught today. He hypothesized, based on his observations, that there was a factor that transferred traits from parents to offspring (Mendel, 1866). In 1889, Hugo de Vries coined the term 'pangene' (or 'pangen') for this factor, and in 1909 Wilhelm Johannsen shortened this to *gene*. During discussions in this era of science, all manner of properties were considered to be inherited – including physical traits, behavioral traits, and derived properties such as intelligence or even criminal tendency. In 1953, Hershey and Chase demonstrated that it was DNA that contained genetic information, and not proteins, as was previously believed (Brock, 1990), and shortly afterwards the helical structure of DNA was discovered. There was much excitement in the scientific community, as it seemed that we had learned everything there was to know about inheritance – traits were encoded by genes in the DNA molecule, and this was inherited by the offspring.

However, something rather vital was missed out during this process of the development of ideas of genetics, specifically, it was assumed that everything that was discussed as an inherited trait *prior* to the discovery of DNA as the means of transmitting genetic information, was indeed transmitted via small sections of DNA. In the case of behavior, this view persisted even in the total absence of any supporting data. As Stuart A. Newman has noted (2009), biologists interested in evolution became "hooked on genes" during the early years of the field. This was, he suggests, an "understandable consequence" of the way the biological and physical sciences developed, and the unevenness between these fields. However, we now know considerably more about how living beings are produced, and how traits are inherited, and as a result "there is no excuse not to move on."

Genetics is a complex subject, but we only need to understand a few simple aspects to appreciate the way the confusions surrounding genes and behavior have developed. A gene is a collection of chemicals (known as nucleotides) that collectively contain a pattern that does one of two different things: either the gene is a blueprint for a protein, or it produces a *functional chain* of RNA that regulates genes (i.e. turns them on or off) or otherwise affects the production of proteins. Modern research has complicated this view by showing that a single gene can actually specify more than one protein, and the DNA specifying a protein need not come from consecutive sequences, but these points are irrelevant to this discussion.

Let me reiterate this point, as it is crucial: a gene codes for a protein, or it affects the production of proteins. *There is nothing else we currently know of that a gene does.* Now as it happens, by regulating the production of proteins, genes can and do serve a role in determining the biology of all living things. The structure of your brain, for instance, or the shape of the wings of the swift, are both consequences of particular patterns of genes. The genes do not, however, serve as a perfect blueprint, since environment and other circumstances play a key role in how specific organisms come to be the way they are. As Lewontin (1995) has pointed out, seedlings with identical DNA develop into entirely different plants according to the prevailing conditions where they are grown.

Why, then, is there such frequent talk of a genetic basis of behavior? All too often it is because of a widespread confusion as to the extent to which individual genes can be understood as having specific functions. Consider the well-publicized report published in 1993 in the journal *Science* recounting the work of Dean Hamer and claiming to have found a gene which correlated with homosexuality (Hamer et al, 1993). The media trumpeted this research as having found "the gay gene". This conclusion was later shown to be flawed – a study of identical twins demon-strated that homosexuality was not expressed in both twins, which was problematic if being gay was to be linked to a gene since identical twins are supposed to have identical genes. Furthermore, studies demonstrated that adoptive brothers show greater incidence of 'shared' homosexuality than non-twin biological brothers. This is extremely strong evidence that homosexuality is *not* primarily genetic in basis, although this certainly doesn't mean genes are irrelevant to the story of sexuality. As is often the case when it comes to the relationship between behavior and biology, nature and nurture both have a role to play.

Part of the enthusiasm for genetic explanations of behavior is the consequence of the widespread evidence that behavior can be *inherited*. It is certainly the case, for instance, that 'behavior breeds true' – consider the behavior of specific dog breeds, such as the retrieval instinct of my Labrador, for instance. There are also behavioral disorders that result from problems with specific genes, much as physical damage to the brain can affect behavior. Changes to genes can alter behavior, but this should not be understood as suggesting that genes *cause* behavior, only that they affect it.

Virtually all behavior can be *influenced* by genes, but there is no evidence that behavior is *determined* by genes. There is no gay gene, no gene for intelligence, no gene for violence, no gene for reliability... there is no gene for any behavior, and neither does it seem likely that any such genes will be found. Errors in genetic code can cause specific medical conditions (which have behaviors associated with them), as with Down's syndrome for instance, but that is as far as the body of research currently goes. In this regard, the journal *Science* published an article by Charles Mann (1994) entitled *Genes and Behavior* that contains this fitting quote:

> Time and time again, scientists have claimed that particular genes or chromosomal regions are associated with behavioral traits, only to withdraw their findings when they were not replicated. "Unfortunately," says Yale's [Dr. Joel] Gelernter, "it's hard to come up with many findings linking specific genes to complex human behaviors that have been repli-cated."... All were announced with great fanfare; all were greeted unskeptically in the popular press; all are now in disrepute.

Unfortunately for Dawkins, his imaginative imagery of animals as effectively robots that are programmed by their genes depends upon making the connection between genetics and behavior – and this link has not panned out in practice. This aspect of the selfish gene myth does not match up with the science, and must either be significantly revised or discarded entirely. The cat and the hamster were not programmed by their genes in any meaningful sense, even though their instincts emerge from a process that involves genetics. One obvious option for recasting this story in a new mould would be to weaken the case from genes being the drivers of the vehicles to having the genes be merely the designers – the genes *made* the cat and the hamster, even if they didn't *program* them. If genes cannot be made to be the causes of selfish behavior, perhaps they can be made out to *influence* selfishness.

Myths of Selfishness

There is, alas, a terrible confusion over terms when it comes to discussions of selfishness, largely because in sociobiology the terms 'altruism' and 'selfishness' were unwisely chosen as technical terms despite the common usage of these words having wildly different meanings. As John F. Haught (1995) has observed, whenever science borrows everyday language, it "gives ordinary words a new twist, and this can cause a great deal of confusion". We have to be clear about what a particular term is meant to mean in its specific context – and even then, we won't necessarily eliminate all of the potential problems.

This issue has been meticulously explored by the philosopher Elliott Sober (1988), who demonstrates that the way 'selfish' and 'altruistic' are deployed by evolutionary biologists is radically different from the way these terms are used colloquially. He uses the example of giving someone a piano: this can be altruistic in the vernacular sense (depending upon the motive behind the gift), but in the sense used in evolutionary theory, the piano may distract you from having babies, thus reducing your 'evolutionary fitness'.

Midgley (1985) has raised serious issues about the confusions over motive resulting from the use of 'selfishness' and 'altruism' to talk about genetic inheritance. She notes that the official story is that sociobiologists are not in any way talking about motives when they use these words, which in ordinary contexts imply exactly that. However, the ambiguity that results from deploying such overloaded terminology "makes possible a chronic, pervasive play upon words" that allows a "colorful, familiar psychological myth" to be conveyed illicitly, which Midgley suggests leads in turn to "bad science".

Officially, 'selfish' in sociobiology (and similarly in Dawkins' selfish gene metaphor) refers to an abstract and utterly unfamiliar causal property, namely a tendency towards maximizing the number of gene variants that match a creature's own genetic pattern in future generations. Confusion over wider implications almost inevitably follows. As Midgley half-jokingly suggests, chosing 'selfish' to describe this particularly property is like using 'cruelty' to describe anything that causes suffering to anyone else in a future generation, or 'sloth' to describe every-thing that will *fail* to affect them. She notes: "Why such a word should ever have been chosen, if no reference to real selfishness was meant, is hard to imagine."

A few examples will serve to demonstrate that the choice of word was not accidental, and that proponents of the selfish gene myth actually do believe that gene propagation and selfishness belong together as concepts. David Barash (1980), for instance, states: "Parental love itself is but an evolutionary strategy whereby genes replicate themselves", suggesting that an analysis of the behavior of parents reveals "the underlying selfishness of our behavior to others, even our own children." Richard Alexander (1987) proposes that "despite our intuitions, there is not a shred of evidence" supporting the idea that people act to help other people at their own expense. Michael Ghiselin (1974) presents an even stronger view, suggesting that for any animal "given a full chance to act in his own interest, nothing but expediency will restrain him from brutalizing, from maiming, from murdering – his brother, his mate, his parent, or his child." He adds: "Scratch an 'altruist', and watch a 'hypocrite' bleed."

As biologist David Sloan Wilson (2009a) observes, each of

these authors writes as if evolutionary theory "ratifies the concept of *individual self-interest* as a grand explanatory principle." He also notes, just in case anyone might think that these were books written for a popular audience that might be pursuing a Dawkinsian strategy of hyperbolic imagery deployed for dramatic effect, that all three of the books from which these comments were taken were serious academic books, written for other scientists to read. He calls Dawkins' selfish gene a "semantic confusion" (Wilson, 2009b), albeit principally as a result of issues aside from the conflation of selfishness with gene propagation. Midgley (2003) comments that because the language of motive is a natural and habitual way of talking, authors such as those quoted above, constantly confuse their technical terms with their everyday meanings, and end up "supposing that they have radically explained human psychology" rather than having fallen into a philosophical quagmire.

Part of the key problem, as Wilson, Midgley and Sober all recognize, is mistaken beliefs about selfishness in the everyday sense of the word. In philosophy, the belief that all behavior is fundamentally selfish is known by the term *psychological egoism*, which Sober (2000b) suggests amounts to a claim that "all of our ultimate desires are self-directed", and since we don't need to consider other related accounts with similar titles we can simply refer to this thesis as *egoism*. According to egoism, when we want others to succeed the only reason we do so is that we believe that their success will have some beneficial consequence for our own welfare. Egoism has had a significant influence in social science, and even non-economists will often claim that they help others "because this makes them feel good about themselves, or because they seek the approval of third parties."

Sober notes that a standard objection that philosophers offer to egoism is that it is not a testable hypothesis, although he suggests that this criticism is essentially flawed. According to

Sober, we cannot know that some future theory might be developed that would allow egoism to be tested, and thus suggests that we can't object to something as untestable because we don't have an "omniscient grasp of the future of science." This strikes me as a very strange kind of complaint to make: Sober seems to recognize that our current circumstances render the idea of testing egoism empirically to be beyond Popper's milestone, but wants to leave open the possibility that it might later cross the border into empirical testing. He may be right, but so what? While it is out there in metaphysics, the possibility that it might later become science is neither here nor there. If we want to allow the chance that some metaphysical ambiguity might later become empirically valid as a reason to exclude it from metaphysics, we might just as well discard Popper's milestone entirely.

A more relevant objection that Sober offers is that if egoism is untestable, then so are its alternatives. If the case for rejecting egoism is that it cannot be tested, then alternative views in which motives transpire to have more variety will equally prove to be beyond Popper's milestone. This is broadly my view – attempts to explain motivation in terms of grand sweeping theories like altruism (or more varied alternative approaches) just aren't going to offer up any kind of viable empirical test, at least not within the paradigms of science we're currently using. But for me, this is *still* an argument against egoism since I can find no reason we should need or expect to find a single overarching explanation for motive. Rejecting egoism does not compel us to accept what Sober calls 'motivational pluralism' – we could refuse to play this game altogether and simply discuss motives on a case-by-case basis, without any attempt to position them as part of a wider belief about *all* motivations.

Despite these sensible complaints, Sober mounts an argument based on evolutionary principles in order to defend the idea that egoism is unlikely to be a true account of our motives. He considers the plausibility of thinking about parental care from the perspective of a special kind of egoism, namely *hedonism* (or psychological hedonism), the idea that the only ultimate desires individuals have are seeking pleasure and avoiding pain. He suggests that, from a purely pragmatic perspective, hedonism isn't a very reliable mechanism for delivering parental care – if the only reason the parent acts is to lessen their own unpleasant feelings, there will be many situations in which other competing fears and anxieties might make helping a child undesirable. An altruistic explanation of parental care doesn't have this problem – in such a case, the care of the child is a *direct* motivation, and as such will be more reliable.

Sober notes that, from the point of view of evolution, "hedonism is a very bizarre motivational mechanism". From the perspective of natural selection, what matters is a creature's ability to survive and reproduce, and in this respect what matters is the survival of the animal itself *and* its offspring. But hedonism proposes that creatures care solely about their states of consciousness, which is rather peripheral to their survival and reproductive success. As Sober suggests: "Why would natural selection have led organisms to care about something that is peripheral to fitness, rather than have them set their eyes on the prize?" It is certainly the case that avoiding pain and seeking pleasure are two of the major goals that we and other animals pursue – but mammals and other creatures of similar complexity form complex representations of the world around them, and as such we can hardly be surprised that we care about more than just these raw sensations.

Thus Sober concludes that a purely egoistic set of motives is less likely to have evolved than a set of motives including both egoistic and altruistic ultimate desires. He does not suggest this proves that motivational pluralism is true, but he claims that his argument at least demonstrates that treating egoism as the default hypothesis for motive lacks plausibility. While I agree in broad strokes, I find it strange that anyone would feel it necessary to turn to an argument in terms of natural selection in order to resolve this question. The reason this problem occurs isn't because egoism is a particularly plausible theory, nor because it can be empirically validated, but rather that a great many people are strangely skeptical of altruism. Recall Ghiselin's odd claim that all altruists are hypocrites.

This kind of faith in egoism has become endemic in contemporary culture, and can be seen historically as a development of the Enlightenment focus on the individual, itself descended in part from Thomas Hobbes view that co-operation only occurs as a consequence of enlightened self-interest. On Hobbes' view, if it were not for governments imposing a need to co-operate life would be a "war of all against all", and thus people consent to a social contract in order to avoid this frightening "state of nature" (Hobbes, 1651). Hobbes *Leviathan* transpired to be a highly influential work of political philosophy, one that led many to conclude that self-interest was the only motive in any interactions that people undertake, that is, that egoism is the only legitimate theory of motivation.

But as Midgley (1985) has pointed out, this is a bizarre story since if this claim were true then the notion of selfishness could never have arisen in the first place: if it was impossible for individuals to care for other people there could not have been a word for failing to do so. Midgley points out that the way we use 'selfish' in its everyday sense is inherently negative in its connotation – it expressly means a deficit of the normal regard for other people. Calling someone selfish *doesn't* mean that they are prudent or self-preserving, it expressly means that they are exceptional in not caring about other people – they are, in effect, broken because they lack empathy, and having feelings for other people is a normal part of being human.

As it happens, empathy is also part of being a mammal, and perhaps even a bird, at least to some extent. Hobbes' idea of the "state of nature" being a vicious competitive war between all individuals reflects the intense prejudice against animals that has persisted from previous centuries. In her first book, *Beast and Man*, Midgley (1978) explored the range of prejudices against animals that are deeply embedded in Western conceptions of 'beasts'. It is no accident that 'beastly' is a pejorative. She notes that we think ill of the wolf, for instance, because we picture her as she appears to the shepherd at the moment of stealing a lamb from the fold. Yet this is like judging the shepherd by the moment he finally decides to turn the young sheep into lamb chops.

Despite Hobbes' beliefs about nature, Midgley notes that aggression among animals is rarely aimed at killing – it more commonly aims at submission, or at driving an interloper out of private living space. In the case of the cat and the hamster, the cat's aggression is even less bound up in hostility – it essentially *plays* with the rodent, even if the experience is a long way from fun for the hamster! Of course, predators do eat meat, and this involves killing. Nonetheless, we wildly misunderstand the nature of life if we mistake the extremity of the culmination of the chase between predator and prey as the entirety of the natural world. Hobbes image of a "war of all against all" as the state of nature implies just such a violent circumstance as the permanent state of life for animals, a situation held at bay among humans only by their consent to being governed. Few naturalists would accept such a narrow view of animal motive.

In the opening section from *The Selfish Gene* quoted above, Dawkins claims that we can expect genes to be "ruthlessly selfish" and that we are "born selfish". Tellingly, he also suggests that if we want a society in which we co-operate, we must teach altruism because we "can expect little help from biological nature". These claims reflect Hobbes' strangely narrow view of the natural world, but they cannot be correct. Our empathy, and the similar capacity for care found in many other animal species, is clearly something we have inherited from nature. We are not "born selfish", as Dawkins suggests, although certainly one of the problems that parents face is preventing their children from *becoming* selfish, that is, becoming hooked on their own satisfaction at the expense of the welfare of those around them. Selfishness is something anyone *can* become, but it is not anyone's default state.

By taking too seriously Hobbes' narrow view of motive, we create imaginary problems that do not exist. The 'problem of altruism', as Midgley (1985) has observed, keeps recurring. Rather than accepting that altruistic behavior is as natural as self-interested behavior (and exists in natural tension with it), myths are offered in which organisms are understood as actually pursuing what is purportedly their own advantage, in terms of having as many descendents as possible. Or alternatively they are recognized as behaving altruistically but only as "the dupes of their genes", which are offered – rather oddly – as conscious agents, "egoists behind the scenes organizing the performance."

Dawkins' myth of the selfish gene cannot be made viable – it overinflates the importance of genetics and simultaneously implies an egoism that is simply untenable as a scientific belief. Genes do not have motives, selfish or otherwise, and selfishness is not a reasonable description of the default behavior of any animal – even substituting 'self-interest' for 'selfishness' only reduces the inaccuracy of this description. However, the selfish gene story is by no means the only imaginative metaphor that Dawkins has constructed to help explicate evolution for a popular audience, and perhaps we can find something else of use in his metaphorical toolbox that might serve as the basis for a replacement myth.

In 1996, Dawkins published *Climbing Mount Improbable*, in which he expanded on Sewall Wright's concept of an 'adaptive landscape'. Wright's image was, Ruse suggests (2003b) a "powerful metaphor" for understanding how organisms develop – successful creatures can be imagined atop peaks in the landscape, and failures positioned down in the valleys. Evolution in this perspective is a matter of moving from one peak to another. Outside of the mountainous regions of success, the precipitous fall guarantees the failure of any animal that falls off an adaptive cliff, while over successive generations lineages of creatures can advance to ever higher peaks of adaptation.

Dawkins, expanding Sewall's imagery, suggests that the evolution of life could be understood as an ascent up the slopes of "Mount Improbable" – the landscape of all possible life – and although more recent research brings into question just how unlikely this progression might be (Lynch, 2007), Dawkins did hit upon an excellent image for why evolution can seem to attain progress over time. Instead of moving freely about Sewall's 'adaptive landscape', there is a sense in which successful features are protected by a ratchet, which prevents them from falling back down the slope, since having found a successful peak in the landscape an animal's descendents either remain on the high ground or die out. Dawkins sees this ratchet as the key to why evolving organisms are able to accumulate functional benefits.

Rather than the selfish gene myth – which risks misrepresenting the natural history of life as driven solely by miserly competition (a fallacy exposed in the last chapter) – the gene's eye view can be illuminated by the alternative myth that *advantages persist*. This is indeed the essence of the gene-centric view: a gene that leads to advantages for an individual (and by extension, a species) is vastly more likely to persist, and this persistence of benefit is the ratcheting mechanism that preserves the capabilities of those creatures that have flourished in the past. Advantages persist because the chain of inheritance carried them forward from the past since, as Dawkins himself has said (Brockman, 1995), "the world is full of genes that have come down through an unbroken line of successful ancestors, because if they were unsuccessful they wouldn't be ancestors and the genes wouldn't still be here."

Why Co-operation?

Although we can put aside 'selfish' as a meaningful biological assertion, the question of *why* creatures co-operate to the extent that they risk their own lives does present itself as something of a dilemma when the evolution of life is approached from the gene's eye view. Looking solely from the perspective of genes invites the assumption that there is a problem with any behaviors that act against the propagation of a particular gene variant – if a behavior works against a gene being passed on, how could the genes that facilitate this behavior have originated and why would they have persisted?

It's important to appreciate that in the orthodox gene's eye view, co-operation isn't strictly problematic. If two animals co-operate for mutual benefit, so much the better for both their genes. The assumption that the inherent background of competition will block co-operation is misguided, and not just because the importance of competition in the history of life is frequently overstated. Taking into account only self-interest, it is often possible to gain additional benefit from co-operating: cleaner fish of various kinds (for instance, remora that attach themselves to sharks) gain nutrition from scrubbing bacteria and algae from the skin of other fish, while their "clients" attain better health by having dead skin and infectious agents removed. Since everyone benefits, this kind of co-operation is unproblematic from the gene's eye view.

The problem comes when facing situations where one animal sacrifices its chance to have offspring or risks its life for another. This looks (at first glance) to be ruled out by gene-centered arguments, yet it happens all the time in nature. As examples of the former, consider social mammals such as meerkats or dwarf mongooses where within any pack usually only one pair breeds; or ants nests, where worker ants are sterile and only the queen has offspring. As examples of the latter, consider alarm calling in squirrels, where the loud noise puts a squirrel in danger of being caught by an attacking predator but helps guard other squirrels nearby; or the white tail of deer and rabbits, whose bobbing motion attracts attention and allows one animal to lead a predator away from the rest of its group.

Possible solutions to this problem percolated through evolutionary biologists working in the mid-twentieth century. J.B.S Haldane developed the basis of an approach, which was later formulized by W.D. Hamilton into what John Maynard Smith termed *kin selection* (Ruse, 2003b). The idea is simple enough: if you die before you have children, your genes die with you. But your relatives share a proportion of your genes, so if you die saving your relatives, your genes can live on. A famous story regarding Haldane says that he was seized with the idea while in a pub, and proceeded to scribble calculations on the back of an old envelope before declaring "I am willing to die for four uncles or eight cousins!" (Midgley, 1985).

More formally, a gene which influenced behavior towards this kind of fraternal sacrifice could propagate itself in a population if the death of one animal carrying that gene helped the survival of other animals carrying the same gene. In what we can call *Haldane's jest*, an uncle has a one in four chance of carrying the same ancestral gene variant as you, while a cousin has a one in eight, hence the numbers quoted above. Hamilton's mathematical version of kin selection had at its heart a simple mathematical inequality, known as *Hamilton's rule*, such that a gene encouraging self-sacrifice can propagate if the 'benefits' of this act outweigh the 'costs', with the degree of relatedness weighting the extent of the benefit. Hence, if you are carrying a gene variant that promotes self-sacrifice, that gene can become more widespread if you die saving the life of four uncles or eight cousins, any of whom could be carrying that gene (presuming the costs and benefits involved are equivalent).

By apparently solving an otherwise insoluble problem in evolutionary studies, kin selection became enormously popular among evolutionists, and the term *inclusive fitness* entered into the canon as a description of what natural selection optimizes. The idea here is that rather than counting solely a creature's offspring when imagining fitness, relatives can also be counted too, modified by a fractional value representing relatedness. Crudely, therefore, if a hypothetical animal is survived by two children, but also eight nieces and nephews, its inclusive fitness will be proportional to three (two plus eight-times-one-eighth), rather than two. Remember that 'fitness' here is only a metaphorical measure of reproductive success, and 'inclusive fitness' is similarly only metaphorical, but these abstractions can still produce hypothetical claims when they are found in equations such as Hamilton's rule. Elliott Sober and David Sloan Wilson (2011) explain this point clearly:

> Before inclusive fitness came along, it was natural to think about individual selection by imagining that individuals "try" to maximize their Darwinian fitness. Although "trying" can't be taken literally, the *as-if* quality of this thought is often heuristically useful; we often can predict which traits will evolve by imagining rational agents who are trying to get what they want... Inclusive fitness seems like a natural generalization of this idea – individuals are "trying" to maximize the representation of their genes in future generations, where it is recognized that your genes can be found in your genetic relatives as well as in your own offspring. The idea then gets broadened further, by taking into account the fact that nonrelatives sometimes have copies of your genes (though here "your genes" means genes that are identical by type, not identical by descent); this means that helping nonrelatives can also be a way to get your genes represented in future generations.

Part of the appeal of the kin selection approach was that it enabled altruistic behaviors to be interpreted as a form of self-interest, because the animal making a sacrifice for its relatives maximizes its own inclusive fitness by helping "its" genes. David Sloan Wilson notes that inclusive fitness "made evolution seem just like economics, in which everything can be explained as a form of utility maximization at the individual level." (Wilson 2009c). Indeed, Hamilton's rule is *only* this – an optimality model. It is not really a model that explains how the behavior in question might evolve at all, and indeed, how it might be practically applied "remains a bit of a mystery" (Jensen, 2010).

Thinking in terms of kin selection has a subtle hidden cost: by envisioning *any* form of co-operation that evolves by natural selection as a form of genetic self-interest, the idea of 'self-interest' becomes "an all-encompassing category" (Sober and Wilson, 2011). Philosophers are naturally suspicious of ways of thinking that achieve totality via the way they are defined – the satisfaction a theorist can feel at having devised an apparently universal theorem can mask a severely limited perspective. As we have already seen, the idea that self-interest can serve as an ultimate explanation for behavior (i.e. egoism) goes wildly beyond Popper's milestone, and can only reasonably be considered metaphysics. It is for this reason that kin selection, while legitimate science in one sense, also becomes mythic, an imaginative story that guides thought towards specific conclusions, and away from alternative perspectives.

In the case of kin selection, the perspective that became brushed under the carpet was *group selection*, the distinction here being between selection in the context of explicit relatives (kin selection) and selection in the context of sets of animals *irrespective* of whether they are related (group selection). The evolutionists responsible for developing kin selection were aware that there was a possibility that selection might occur at the level of the group rather than the level of the individual, but the consensus – especially in the light of the formulation of inclusive fitness and kin selection – was that it was too weak a force to have any significance. John Maynard Smith (1964), for instance, was willing to accede the possibility that group selection *might* occur, but felt that the necessary conditions for it were unlikely to come about in practice. The party line became that the only kinds of groups that mattered in the context of selection were genetic relatives, and that any kind of non-related groups of creatures were unimportant to evolution.

George C. Williams, an evolutionary biologist specializing in fish, was a particularly influential voice in the case mounted against group selection. His 1966 book *Adaptation and Natural Selection* laid out the basis for the gene-centered view of evolution (and was, in fact, a major source of the ideas Dawkins presented in *The Selfish Gene* ten years later). Williams examined a variety of possible cases of group selection (mostly in fish species) and determined in each that an explanation in terms of individual selection was always available. He thus concluded that selection between groups was "impotent in a world dominated by genic selection and random evolutionary processes". He did not dismiss the possibility of group selection completely, but asserted that biologists should "postulate adaptation at no higher a level than is necessitated by the facts." Williams was mindful of the inherently changeable world of scientific models, however, admitting that there was still much to be learned: "I am sure that by the standards of a generation hence, our current picture of evolutionary adaptation is, at best, oversimplified and naïve."

Sober and Wilson (2011) recently returned to Williams' seminal text in order to reassess the relative merits of the kin selection and group selection perspectives. Praising much of the foundational work Williams pursued, they draw particular attention to the idea that "adaptation at a level requires selection at that level", calling this *Williams' principle*. This adage sets the requirement that must be met for group selection to be considered valid: the fact that some trait provides a recognizable benefit to groups isn't enough for that trait to be considered a group adaptation – it must have developed precisely *because it was beneficial to the group*. Using this as the basis for their discussions, they criticize a number of biologists for violating Williams' principle, and thus continuing to defend a kin selection-inspired perspective against group selection.

The biologists in question are typical examples of the exaltation of genes this chapter addresses, maintaining (Sober and Wilson assert) "that the individual is *always* a unit of adaptation no matter what the mix is of group and individual selection." Sober and Wilson point out that this position is being held on the basis that individuals are predicted to maximize their inclusive fitness, and this claim holds true irrespective of the relative strengths of individual or group selection. Yet this violates Williams' principle. Kin selection arguments can't rule out group selection from happening, since inclusive fitness only entails claims about the rates of gene survival – it has nothing whatsoever to say about the *scale* at which adaptations occur. In fact, as philosopher of science Samir Okasha (2004) has noted, rather than claiming individual selection always occurs, there are many situations in which the question of whether or not there is individual selection taking place is not defined.

There are now a vanguard of scientists and philosophers who recognize that it was a mistake to believe that group selection is a force too weak to have any influence in evolution, and champion what is called *multi-level selection* – the idea that selection and adaptation can occur at the level of the individual, the group, or even at the level of an ecology. Note, however, that in all these cases, the accumulation of changes *still* happens via alterations in gene frequencies (although it is important to remember that even among animals there are also persistent cultural and environmental effects that play an important role in the story of life). The breakthrough realization has been that the metaphorical 'selection' that takes place occurs within a specific context. It is always possible to render these effects solely in individualistic terms, but transformable perspectives are not always equivalent in explanatory value. If group selection played a major role in the occurrence of a particular trait, is it really helpful to talk about that trait as an individual adaptation?

This is not just, as Dawkins (2007) has accused, merely "semantic doubletalk". Empirical results validate the multi-level selection perspective, and show how focusing *solely* on genes can only give an incomplete perspective on evolution. For instance, William Muir (1996) has demonstrated that artificial group selection of hens results in decreased mortality of the birds, and increased egg production, with the bulk of the improvement occurring within just three generations. As another example, William Swenson, David Sloan Wilson and Robert Elias (2000) showed that artificial group selection of microbial ecosystems in soils led to improved growth for plants grown in that soil. Far from group selection being too weak a force to have any serious influence, it seems it is strong enough to have dramatic and demonstrable effects.

However, if Kuhn taught us anything about the history of science it is that research communities do not change their beliefs without a fight – often, you might say, to the death. Hamilton, however, who did much of the heavy lifting on kin selection and had been originally opposed to the viability of group selection, changed his view almost completely as a result of a fascinating equation by George R. Price that, alas, is slightly too complex to explain here. As Wilson (2009c), notes, Hamilton was happy to change his mind once he recognized that "inclusive fitness theory is not an alternative to group selection after all; the role of group selection was merely obscured by the way it was formulated."

What really put the cat among the pigeons in recent years was a paper by Martin Nowak, Corina Tarnita and Edward O. Wilson

(2010) that denies the validity of using kin selection to explain the origins of the behavior of social insects such as ants. Edward O. Wilson is particularly famous in evolutionary circles as the founder of sociobiology, not to mention an expert on insect species of all kinds, and his apparent change of heart in respect to the validity of group selection raised both eyebrows and tempers. In fact, he has been moving in this direction for some time, and published several earlier papers dealing with the topic (for instance, E. O. Wilson, 2008), all stating the same thing: kin selection wasn't a significant factor in the evolution of the social behavior of ants, while selection between colonies (i.e. multi-level selection at the level of the colony) appears to have played a highly significant role.

Ants and wasps had generally been considered a great example of the merits of the inclusive fitness perspective advanced in kin selection, owing to a rather unique system of inheritance. Whereas most animals pass on half of their genes to their offspring (hence children share half their genes with each of their parents, and on average half their genes with each of their siblings), many social insect species have females with genes from both parents but males with genes solely from the mother (because the males hatch from unfertilized eggs). The male ants always pass on the same genes, thus when a queen mates her daughters get *all the same genes* from the father, plus half of the genes from their mother – meaning all female ants in a colony have *three quarters* of their genes in common with each other. In terms of Haldane's jest, this means that three female ants would be willing to sacrifice themselves for four of their sisters – a pretty good rate of exchange! These numbers were sometimes claimed to be a validation of inclusive fitness, serving to explain why ants and bees had such tight social structures.

The trouble is this explanation never actually stacked up in practice. For a start, remember that the formula for kin selection, Hamilton's rule, is only a description of optimal conditions – it isn't actually an explanatory mechanism. As Christopher Jensen (2010) has noted, it's rather remarkable that no-one has incorporated Hamilton's ideas into a predictive scientific model that demonstrates how these insect species reach the optimal state predicted. In the absence of such a model, as Nowak and colleagues suggest, the high relatedness might be better understood as a consequence of the highly social living arrangements, not as a cause.

What's more, as Raghavendra Gadagkar (2010) reports, the actual level of relatedness in insect colonies is considerably lower than the ideal values suggest, not to mention there are a wide variety of social insects that don't have the unusual inheritance pattern described above. Gadagkar suggests that an exclusive focus on relatedness isn't good enough for proving Hamilton's rule; the benefits and costs have to be measured in some way, not simply brushed under the carpet. His own research group has attempted to find ways to attend to this deficit, and when they have done so they have concluded that "ecological, demographic and physiological factors" have a bigger role in bringing about the unique social arrangements of communal insects.

Commenting on the controversial Nowack, Tarnita, and Wilson paper, both Jensen and Gadagkar reach similar conclusions: kin selection and group selection aren't really in competition with each other, they are simply different tools that need to be applied in different ways. Jensen notes in particular that there are situations in which Hamilton's rule can be useful – it just shouldn't be taken as the default explanation. Furthermore, even if group selection is a better explanatory device in respect of social insects, "we need to acknowledge that different theoretical constructs will be useful to explain different evolutionary phenomena."

Regarding the intensity of the backlash against the paper, Jensen recognizes that there are evolutionists who are "wholly committed to the idea of gene-level selection" and who "insist that all arguments rest on the tenet that the gene is the sole unit of selection". He comments that the paper has thus "exposed some very disturbing things about the sociology of evolutionary biology":

This publication has really angered a great number of people, and this anger has revealed the kind of tribal orthodoxy that exists in the mainstream of evolutionary biology. A great number of evolutionary biologists will become enraged whenever someone in the field espouses a hypothesis that contradicts the gene-centered approach to explaining evolutionary processes... the anger often seems excessive in relation to the affront. After all, if an idea is clearly dumb and can easily be dispatched by theoretical or empirical demonstration, why get upset about it? We need more publication of relevant results, not more rhetoric.

It is worth reiterating that the gene's eye view is a perfectly workable scientific metaphor. It only creates problems when it develops into a monomaniacal fixation with genes. Yes, genes are absolutely crucial to understanding the evolution of life, because without changes in the abundance of different gene variants much of what we consider natural selection simply has too little to work with. But when it comes to the evolution of co-operation, the kin selection perspective has radically less to offer than multi-level selection theory.

Co-operation allows groups of animals – even unrelated animals – to gain a significant edge over rival groups that don't co-operate, or who co-operate less well. This situation can lead to the evolution of group adaptations for co-operation, as appears to have happened with social insects. And of course, even without group selection for collaboration, co-operation can *still* be the best strategy. Rather than the myth that collaboration is constrained to occur only between relatives, we need a new myth that expresses the inherent advantages of working together. Rather than kin selection, perhaps we should recognize that *cooperation is an advantage.*

Evolving Alliances

One of the most significant changes to our views of the nature of life in the last century has been development of the concept of an *ecosystem*, a term credited to the British biologist Arthur Roy Clapham, who first started using it in the 1930s (Willis, 1997). Broadly speaking, an ecosystem describes all the living things in a particular region, as well as all the non-living things that interrelate with them. Thus a rainforest ecosystem includes all the trees, the animals that live in them, as well as the soil, rocks, water, air and sunlight that are necessary to maintain it. Midgley (2003) notes that her 1971 copy of the Oxford English Dictionary doesn't include the word 'ecosystem', yet it has become an important part of our thoughts about nature. Thinking about ecosystems stresses the relationship between species – a relationship that has many more dimensions than competition alone.

One of the basic problems with kin selection is that it embeds the kind of Spencerist bias exposed in the previous chapter While competition is a part of life, it is never the whole of the story, and focusing exclusively upon competitive interactions presents a caricature of nature. The concept of an ecosystem helps provide some desperately needed color to our image of how life functions, since no creature can actually be understood in isolation. Every animal depends on others in many different ways – there is much more going on in any given environment than eating and breeding. The ecosystem imagery is also closer to Darwin's fitness to environment metaphor, and clarifies it: the niche a creature fills is not just the product of a sterile landscape, it is also a consequence of the other forms of life that dwell in any given place.

A canopy tree in a rainforest, for instance, can be thought of as in competition with other similar trees for space and light, and in competition with certain insects and animals that eat its leaves or otherwise whittle its resources. However, the tree's very existence creates a rich new environment that many animals live in, without any significant aspect of competition in effect. In fact, many insects and other animals are vital to the trees, being necessary for pollination and the spreading of seeds. The arrangement is generally equitable – a pollinator receives nectar, the flesh of a fruit pays for a ticket for a seed to ride in a digestive tract and end up further afield than it could ever do otherwise. It has been estimated that each species of tree in a rainforest typically has thirty species of insect dependent upon it, while in turn each tree is dependent upon certain animal species for its life cycle. If a tree uses a bat for pollination and a bird to disperse its seeds, the loss of either species can lead to the death of that species of tree in any given region (Butler, 2006).

Of course, nothing about this inter-dependence between species is a challenge to the gene's eye view – Dawkins (1996) has written at length about the close relationship between fig trees and fig wasps, for instance. There are some nine hundred species of fig trees, and nine hundred species of fig wasps that transport pollen between trees of the same species. From the gene-centric view, this situation is perfectly explicable as a co-evolution between two species, both of whom gain benefits from their arrangement. However, notice that the kin selection myth is radically inadequate to explain situations such as these where the genetic make-up of the co-operating species is radically distinct. Fig trees and fig wasps are not even remotely close relatives. Nonetheless, the benefits secured by this inter-species co-operation endure, because it is part of the nature of life that all advantages tend to persist.

When we think of advantages persisting instead of selfish

genes, it becomes clearer that co-operation will not be eliminated by natural selection – it will in fact prosper whenever it is advantageous. As Peter Kropotkin observed in the 1890s, co-operation and mutual aid contribute significantly to the survival of species, since a group of animals can achieve more than the individual members can on their own. The rise of humanity as a species appears to have been significantly facilitated by its capacity to act as a group (Kropotkin, 1902). Kropotkin wrote about co-operation precisely because the Spencerist ideology had dominated discussion of evolution in the late nineteenth century. Although he wrote his book in exile in England, he writes extensively about his scientific expeditions into Siberia, reporting widespread evidence of cooperation among non-human animals, and concluding that mutual aid has been vital to the survival of a great many species.

Although Kropotkin mostly focused on cooperation between animals, it can be found extending into the world of plants and beyond. Plant ecologist Elizabeth Arnold has studied the evolution of plant-fungal co-operation, and together with a team of researchers discovered a unique kind of fungus known as an endophyte that occupies the tissues of plants without causing any disease or other deleterious side effect. Until recently, these microscopic fungi were inexplicable, despite being entirely ubiquitous in plant leaves. Arnold and her colleagues discovered that by hosting these visiting micro-organisms, tropical trees like the canopy trees of the rainforest can protect themselves from infection by hostile fungi and other pathogens that would serve to threaten the life of the tree (Arnold et al, 2003). Again, the fungus and the plant are very distantly related – we are more closely related to the fungus than to the plant (Shalchian-Tabrizi et al, 2008)! – but the advantages of co-operation still win out.

Throughout the natural world, examples of beneficial co-operation – known as *symbiosis* – can be found again and again. In every environment, there are examples of genetically distinct organisms which co-operate, and thus gain advantages in survival and reproduction. Biologists disagree about the term 'symbiosis', however, with two competing definitions sparring for control of the word: symbiosis as persistent mutual benefit, and symbiosis as any long-term association between different species (Douglas, 2010). Angela Douglas has suggested that the former definition is viable, provided the focus is on the *interaction* between organisms and not the organisms themselves – a beneficial relationship may later turn parasitic, or vice versa. The relationships between creatures are always dynamic, and besides, we frequently misjudge the situation: organisms originally thought of as parasites have turned out to be harmless or even advantageous under certain circumstances.

Symbiosis can be found in every ecosystem, on land and in the sea. Ocean life, which certainly has its savage side, is as much a testament to the extent co-operation affords advantage as anything found above the waves. The coral reef, nature's cosmopolitan cities, contain species such as cleaner wrasse and cleaner shrimp that remove parasites and dead tissue, even from within the mouths of other fish, who do not harm the cleaning animal (Eibl-Eibesfeldt, 1955). More than forty five species of fish are known to serve as cleaners (along with at least six species of shrimp) and innumerable fish species benefit from their services (Trivers, 1971). What's more, fish tend to form a long-term relationship with their cleaners, and often arrive at a particularly time of day to have their parasites picked away (Feder, 1966).

Robert Trivers explains all such behaviors in terms of selection for reciprocal altruism, by applying cost-benefit analyses to situations such as the cleaner fish in a manner not entirely disconnected from Hamilton's rule. Predicating self-interest (as criticized above) Trivers seeks to explain cleaner behavior in terms of *direct* benefits between both parties involved. He correctly concludes that co-operative behavior can come about even between unrelated species, but his description rests on the concept of gene variants that bring about co-operation, and this is problematic. The instinct not to eat the cleaner fish is certainly inherited and not learned – a grouper raised in isolation still has the instinct not to eat cleaner fish (Hediger, 1968) – but explaining this behavior by benefits to specific genes isn't going to be sufficient.

Trivers own discussion of the behavior makes it clear that cleaners have acquired a specific 'livery' – unique signaling behaviors and colorations – to announce their 'job', while many of those fish that benefit have their own specific motions or color changes that are used to request cleaning. If we look beyond the genes, it becomes clear that from the perspective of the ecosystem these symbiotic interactions cannot be adequately explained solely from individual selection. An explanation closer to group selection is required, since what has emerged is not an advantage for a lone species of fish, but an advantage for an *entire coral reef*. The inter-related behavioral traits being explained only make sense in the context of the ecosystem they developed within, particularly when a great many different (unrelated) species display the same traits.

Any animal that acquires the 'livery' of a cleaner can 'work' as a cleaner – genetics isn't as important here as it is in other instances, since while the offspring of individual cleaners will inherit the tools of the trade, new species can 'set up shop' as cleaners just by acquiring the relevant traits. As Paul Griffith and Russell Gray observe (2001), the cleaning behavior specifies *trait-groups* (i.e. sets of creatures sharing a specific trait) who aren't related but who collectively gain advantage, and this is a clear example of group selection. Indeed, David Sloan Wilson specifically coined the term 'trait group' (Wilson, 1975) with the meaning of a group that is affected by every individual within it (Wilson, 1979). As Martin Nowak and Karl Sigmund comment (2007), rather than solely the self-interested 'you scratch my back and I'll scratch yours' scenario Trivers prefers, co-operation can also occur on the basis of a *reliable* interaction: "presumably I will not get scratched if it becomes known that I scratch nobody."

In this regard, Dawkins (1976) talks of a fictitious "green-beard effect", such that people with green beards are more inclined to help others with green beards. According to the story as Dawkins presents it, if you have the gene variant for the green beard you help others who have the same gene. But as Griffith and Gray recognize, having a green beard *doesn't* necessarily mean having the same gene variant – just as having cleaner 'livery' doesn't mean being related to other cleaners. Rather than kin selection being a special case of gene selection, as Dawkins suggests, the green-beard example suggests kin selection *is a special case of group selection* (Griffith and Gray, 2001).

The traits that allow for co-operation with the cleaner fish in coral reefs evolved among groups of unrelated species, as did the 'livery' traits of the cleaners themselves, and only a justification in terms of these trait-groups will be sufficient to explain their evolution. Thus a multi-level selection approach forms the basis of a far more workable explanation for cleaner fish and their 'clients' than focusing solely on the genes. As Griffith and Gray point out: "It is not necessary that individuals who selectively benefit one another be related or that they share a gene" since what is important is that something creates "a statistical association between dispensing benefits and associating with other individuals who dispense benefits." Selection at the level of groups of creatures with common traits requires an understanding of the interactions that occur within their ecosystem, and situations such as these cannot be interpreted solely in terms of individual selection or genetic cost-benefit analysis.

Symbiosis Everywhere

Lynn Margulis, the biologist most famously associated with symbiosis, is explicit about the limitations of the gene's eye view when it comes to understanding co-operation in nature:

"Symbiosis has nothing to do with cost or benefit," she bluntly states, and goes on to add that cost-benefit models "have perverted the science with invidious economic analogies." (Brockman, 1995). We already encountered this problem in the previous chapter in the context of Spencerism and the myths of evolution that descend from Victorian economics. Margulis further charges that evolutionary theory is dominated by "aseptic language" resulting from the use of cost-benefit methods "borrowed from insurance practices" (Margulis, 1991). She raises complaints about the way individuals are defined that mirror those discussed previously in the context of the ambiguous term 'fitness', but her issues are not quite the same as the objections raised by Ariew and Lewontin.

In 1966, Margulis wrote a paper (under her married name of Lynn Sagan) that speculated on the origin of the more complex cells of all multi-cellular organisms, what are known as 'eukaryotic cells' as opposed to the 'prokaryotic cells' of bacteria. The paper was rejected by fifteen scientific journals, not only because it had (Margulis admits) many flaws, but also because it was entirely new and no-one was capable of evaluating it (Brockman, 1995). Eventually it was accepted for publication by *The Journal of Theoretical Biology*. The paper hypothesized that the three fundamental elements of the more complex cell structures were originally separate bacterial cells that merged with other bacterial cells to form a new kind of life. The three examples she gave were mitochondria (the 'power generator' of animal cells), chloroplasts (which turn light into energy in plant cells), and flagella (cellular 'tails, like the tail of a sperm cell). In each case, she proposed how a new form of life came into existence through the merger of two distinct single cellular organisms (Sagan, 1967).

At the time this paper was published Margulis was a complete unknown, yet it received an unprecedented eight hundred requests for reprinting (Brockman, 1995). She developed the idea over the following years and eventually produced a book, *The Origin of Eukaryotic Cells*, which outlined what is termed 'endosymbiotic theory' – the idea that the cells of complex forms of life all originated through the fusion of single celled organisms (Margulis, 1970). Reprinted numerous times, now under the title *Symbiosis in Cell Evolution*, Margulis has called the book her "life's work", a radical theory that not only explicated the origins of the different elements of plant and animal cells, it also explained how this led to cell division and sexual reproduction.

Margulis offers an entirely unique perspective on the origin of complex cells:

It may have started when one sort of squirming bacterium invaded another – seeking food, of course. But certain invasions evolved into truces; associations once ferocious became benign. When swimming bacterial would-be invaders took up residence inside their sluggish hosts, this joining of forces created a new whole that was, in effect, far greater than the sum of its parts: faster swimmers capable of moving large numbers of genes evolved. Some of these newcomers were uniquely competent in the evolutionary struggle. Further bacterial associations were added on, as the modern cell evolved. (Brockman, 1995).

Margulis' work had precedents, but they had been largely ignored. The Russian botanist Konstantin Mereschkowski had proposed what he called 'symbiogenesis' as the origin of chloroplasts at the turn of the twentieth century (Mereschkowski, 1905), while in the United States Ivan Emanuel Wallin proposed 'symbionticism' as the origin of mitochondria (Wallin, 1923). Neither theory was taken seriously until Margulis resurrected their work, and took it further. She claims that the major divisions of multi-cellular life on our planet – plants, fungi,

animals and others besides – all have their origin in endosymbiosis (Bermudes and Margulis, 1987). The diversity of life on
our planet is thus *a result of long-term co-operation.*

Take a moment to really absorb the consequences of this
discovery. You and I, every tree and plant, every animal and bird,
every fish and fungus, everything living, in fact, except microscopic bacteria, is a product of co-operation between ancient
organisms who formed a unique 'merger' that has lasted for
billion years. We tend to think of ourselves as individuals, but
from the perspective of endosymbiosis we are nothing of the
kind – we are an alliance of billions of subtly different cells who
co-operate together in order to function as a viable collective. I
mentioned before that focusing upon competition between
companies and countries ignores the co-operation essential to
any such institution: this is equally true when we consider the
nature of life. The fox and the rabbit may be competing for
survival, but the *cells* of the fox and the *cells* of the rabbit are an
example of co-operation on an incredible scale. A certain degree
of competition is an almost inescapable consequence of multi-
cellular life, with its high energy lifestyle and demand for raw
materials and living space. Yet multi-cellular life is inescapably a
consequence of co-operation. Without that co-operation, we
wouldn't be here at all.

Although she initially had great difficulty having it accepted
by other scientists, Margulis' theory is now widely accepted
among biologists. Niles Eldredge has called it "the grandest idea
in modern biology", philosopher Daniel Dennett suggests it is
"one of the most beautiful ideas I've ever encountered" and
considers Margulis "one of the heroes of twentieth century
biology", while Richard Dawkins has called her theory "one of
the great achievements of twentieth-century evolutionary
biology" and admitted his great admiration for Margulis' theory
(Brockman, 1995). However, Margulis herself suggests that
"most scientists still don't take symbiosis seriously as an evolu-

tionary mechanism" (Brockman, 1995). I have to agree. Despite the praise for endosymbiosis theory, it is generally never brought up in discussions of evolution, or is mentioned solely in passing, as a curious footnote. Explanations in terms of genes are almost always preferred, and unique historical events such as those proposed by Mereschkowski, Wallin and Margulis are sometimes downplayed by an appeal to self-interest in its mythic role as a grand explanatory principle.

In addition to her own achievements in discovering the incredible significance of symbiosis for the diversity of life on our planet, Margulis has also had a key role in recognizing symbiosis on a very different scale. In the early seventies, she was attempting to reconstruct the evolutionary history of dozens of different kinds of bacteria and noting that all such single celled organisms produce gases of various kinds – oxygen, carbon dioxide, nitrogen, ammonia... more than thirty gases in all. Talking to biologists, she was surprised that the prevailing belief was that while oxygen was a biological product, the other gases were presumed to be nothing of the kind. She was told by at least four scientists to "go talk to Lovelock" (Brockman, 1995).

James Lovelock had been scientifically trained, but had enjoyed an eclectic career, including designing a number of scientific instruments for NASA's space exploration program. The Viking program in the 1970s was partly motivated by the desire to discover if Mars could support life, and working in the preparatory stages of the Mars probes, Lovelock became interested in the composition of the Martian atmosphere. If there was life on Mars, he reasoned, it would have to be using and altering the compositions of gases. What he found instead was an atmosphere in a static chemical equilibrium, mostly comprised of carbon dioxide. Even before the Viking probes had visited Mars, Lovelock had concluded that there was no life to be found (Lovelock, 1979).

When Margulis and Lovelock came together, it led to another

revolutionary view of the nature of life – the *Gaia principle*. In essence, just as the life in any given part of the world can be studied as an ecosystem, so the interactions between the ecosystems as a whole "compose a single huge ecosystem at the Earth's surface" (Margulis, 1998). Margulis' work on bacterial gas production dovetailed with Lovelock's planetary atmosphere work, and sparked a whole new perspective on life. Although credited as co-founder of this theoretical framework, the implications of which go far beyond the scope of this book, Margulis downplays her role in its formulation, stating that it took her days just to begin to understand Lovelock's way of thinking. The temptation to interpret his concept as simply adaptations to environments was strong, even for an original thinker such as Margulis. She suggests that her contribution was restricted to simply helping him work out his explanations (Brockman, 1995).

According to Lovelock's account of Gaia, our planet functions as if it were a giant organism or *mega-organism* (Lovelock, 1979). Margulis, however, is resistant to this terminology, since "no organism eats its own waste" (Brockman, 1995), while recognizing that Lovelock's imagery helps to communicate the idea of the Gaia principle to a wider audience. Her view, as noted above, is that Gaia can be understood as an ecosystem or, perhaps, a super-ecosystem. Whichever metaphor one uses, the Gaia principle provides the complementary top-down perspective that goes with the bottom-up view beginning with genetics. If the latter gives us the gene's eye view, the former gives us the *world's eye view*. As Greg Hinkle, one of Margulis' former students remarked: "Gaia is just symbiosis as seen from space" (Margulis, 1998).

Dawkins is one of many scientists to express hostility towards the Gaia principle. He suggests, looking just at the issue of gas production among bacteria, that it cannot be the case that a particular kind of bacteria makes a particular gas "for the good of the world" since any individual bacteria that doesn't put itself out by making that gas must necessarily do better. If the bacteria gains individual benefit by doing so, Dawkins has no problem with the scenario – but then, he notes, the Gaia principle isn't needed for explanation (Brockman, 1995). This objection is reasonable, but as we have already discovered selection is not something that happens solely at the level of the genes or the individual – it can happen at many different scales. All that the Gaia principle requires is that multi-level selection can potentially be influenced by selection at the level of the planet as well as at the ecosystem, which is unproblematic.

Perhaps part of the problem here is in thinking in terms of the symbiotic interactions entailed in the Gaia principle as "for the good of the world". Symbiosis, after all, does not necessarily entail self-sacrifice, and is typically a form of mutual aid. Gaia, in both Lovelock and Margulis' viewpoint, is a way of suggesting that the surface of the Earth is "best regarded as alive", it behaves *as if* it were a physiological system, at least in certain senses (Margulis, 1998). What is entailed by Gaia is not sacrifice for the common good (as Dawkins seems to impute) but self-regulation on a global scale. This can be hard to comprehend from the gene's eye view, but it's far easier to understand as trait-group selection like the cleaner fish in the coral reef. In trait-group selection, any (not necessarily related) group of creatures can gain selective advantages collectively.

In the case of Gaia, different organisms, including but not restricted to bacteria, form cycles around specific chemical compounds. These chemical cycles are trait-groups because the actions of individuals have consequent effects for all other individuals in the group. For instance, the nitrogen cycle involves bacteria who 'fix' the gas into chemicals such as ammonia and other nitrates that are crucial for the life of plants. So vital are these bacteria to plants that a vast number of species have nodules in their roots which host symbiotic bacteria for nitrogen fixing. However, a high concentration of nitrates in the soil renders it toxic. Another kind of bacteria removes the nitrates, turning them back to nitrogen: if nitrate concentration falls too far, the soil becomes infertile, if it climbs too high, it becomes toxic. The plants, in fact, are rather new to this process, which appears to have been established by bacteria some 2.7 billion years ago (Canfield et al, 2010).

The success of each of the different kinds of organism in the modern version of this cycle – nitrogen fixing bacteria, plants and denitrifying bacteria – can be explicated on a model of individual selection, but not completely. Small variations in the ratio of nitrogen fixing to denitrification have significant effects on the composition of atmospheric gases, especially carbon dioxide (Falkowski, 1997). Runaway populations of either kind of bacteria are thus selected against at the level of a trait-group since the inter-relations between the species connected together by this chemical process cause the actions of one set of individuals to have significant effects on the others. The evolution of the nitrogen cycle cannot be rendered in terms of pure individual selection without simply pretending that the inter-related forms of life can be abstracted entirely into the background.

It is precisely these kinds of complex inter-relations at the planetary scale that the Gaia principle addresses, placing them all in the context of regulatory systems that are parallel to the same kind of biological feedback loops that allow individual cells or animals to maintain their equilibrium. The latter are generally amenable to explanations in terms of individual selection because the effects of the feedback are focused on single organisms, but once the interactions occur at the level of the ecosystem or super-ecosystem a different approach is required. Selection on scales beyond the individual reinforce and ultimately establish complex meshes of interacting species of bacteria and other creatures as regulatory mechanisms for maintaining the necessary circumstances for life as we know it – that is the evolutionary origin of the Gaia metaphor. This concept is becoming widely accepted outside of evolutionary biology, but anyone committed to gene supremacy finds the world's eye view very difficult to accept.

From individual cells to the planet as a whole, symbiosis is a vital part of the story of life, and one that is usually downplayed or brushed under the carpet – often by invoking faith in self-interest as an explanatory principle. Symbiosis, as co-operation between different species, is an aspect of the myth that co-operation is an advantage that I propose as an alternative kin selection, but it is important to remember that the advantages in co-operation also occur to a great extent between animals of the *same* species. This doesn't qualify as symbiosis, but it is still a highly significant aspect of the nature of life, one easily obscured by excessive focus on competition. Yes, creatures compete with one another, but they also co-operate, and any attempt to present life as fundamentally about one or the other rather than a subtle blend of both can only be misleading.

In terms of the connection between co-operation and genetics, there is one more observation that has to be made. Although as I have already stated it can be misleading to conflate biological mechanisms with behavior, the hormone and neurotransmitter oxytocin appears to be the constitutive mechanism by which trust operates (Baumgartner et al, 2008). There is thus a sense in which there is a 'gene for trust', at least if we ignore the fact that for oxytocin to function at all requires a complete nervous system that is specified by a great many different genes. Almost all vertebrate species possess a protein chemically similar to oxytocin, the ancestral precursor of which appears to have originally been involved in water regulation (Venkatesh et al., 1997). The two relevant genes are close to each other on the same chromosome, and oxytocin is believed to have originated in a gene duplication event some 500 million years ago (Gimpl and Fahrenholz, 2001).

From amphibians onwards, there has arguably been a trend of increasing capacity for trust between animals – something we are most familiar with in the context of mammals, who form packs and other groups for mutual benefit, but which is also found in the flocking of birds, and even in reptiles and fish. Midgley (1978), drawing on the work of Irenäus Eibl-Eibesfeldt (1970), observes that sociality among primates develops from "extension to other adults of behavior first developed between parents and young – grooming, mouth contact, embracing, protective and submissive gestures, giving food." Sociality can thus be seen as adults developing a mutual relationship "as honorary parents and children", and there is therefore a sense that trust and co-operation can be seen as evolving out of parental care. In this regard, note that oxytocin and other related proteins seem to play a critical role in parenthood, since a large quantity of oxytocin is released during childbirth (van Leengoed et al, 1987, Kendrick, 2004).

The trust between social animals (including humans) who co-operate for mutual benefit is another overt sign of the benefits of co-operation, and in this case it is one we *can* connect to a gene. But that gene, and the protein it specifies, does not lend advantages only to immediate relatives, but creates the possibility of trust even between different species. There was a story in late 2005 of a dog in Seattle who had 'adopted' a baby squirrel. The squirrel had fallen forty feet from a tree along with its sister, who died in the fall. The wounded baby squirrel was taken to a woman who had a reputation for helping injured animals, and arrived at a time when the woman's dog was pregnant. Incredibly, the dog insisted on caring for the squirrel as if it was one of her puppies (Dakss, 2005). As the squirrel grew up, the boisterousness of the litter of puppies eventually forced it to learn independence, and it ultimately left the house for the woods. However, it returned at least once to visit its 'family' (Gilmore, 2005).

The dog-adopts-squirrel story is unusual – it wouldn't have been in the news otherwise – but these kind of inter-species adoptions are far from a rare occurrence. In Kenya, a lioness adopted three baby antelopes in succession (Astill, 2002) and after the 2004 tsunami a 130-year old giant tortoise adopted an orphaned hippopotamus (Hatkoff et al, 2006). In Thailand, it is reportedly common practice for captive tigers to suckle piglets, and for pigs to adopt orphaned tiger cubs (Wigmore, 2007). In Bali, a wild long tailed macaque monkey adopted an abandoned kitten (Young, 2010) while in Massachusetts, a crow similarly adopted a kitten (Faber, 2008). Perhaps strangest of all, in Tokyo a dwarf hamster given to a rat snake as a meal ended up forging a rather odd friendship with the reptile, who now shares its cage with the rodent (Associated Press, 2006). These stories sound bizarre, but as humans we encounter inter-species adoptions on a daily basis without giving them a second thought – we simply dismiss them under the term 'pet'. Consider the story I recounted earlier of how my wife and I inadvertently ended up adopting a hamster, or for that matter our family dog. Even the cat that attempted to maul the hamster in the first place was an inter-species adoptee in my neighborhood.

Mammals are only about 220 million years old (Rose, 2006) but the gene for oxytocin appears to be at least twice this old. Thinking in terms of the selfish gene myth, stories of inter-species adoptions and friendships require rationalization – recall Dawkins suggestion that we "can expect little help from biological nature" when it comes to co-operation. This kind of perspective requires that these cases be dismissed as aberrations: the *real* purpose of oxytocin being to protect the genetic investment of offspring, while these other cases are simply a misfiring of this instinct. Yet in so much as the selfish gene myth dictates that genes serve to foster their own propagation, oxytocin has been hugely successful in its 'selfish' spreading of co-operative behavior. Thinking in terms of the alternative myth that advantages persist, however, it becomes far clearer just why co-operation between and within different animal species is so widespread.

I also wish, rather impishly, to make an observation about the radically different perspectives on the nature of life offered by Dawkins and Margulis – two scientists who, just as with Gould and Dawkins, ended up as bitter 'enemies'. Dawkins' gene's eye view prioritizes competition, which might be connected to the chemical testosterone that has been associated with competitive behavior (Elias, 1981, Booth et al, 1989). Margulis' world's eye view prioritizes co-operation, which might be connected to the chemical oxytocin, associated with maternal care. There's a sense, therefore, in which this particular spat can be seen from the gene's eye view as the two scientists having been 'manipulated' by the rival genes for testosterone and oxytocin into what seems from the world's eye view to be a bizarre 'battle of the sexes' – Dawkins being the stereotypical testosterone-influenced male, and Margulis the stereotypical oxytocin-influenced female, at least in this tabloid version of their disagreement. In fact, both men and women are equally affected by these two chemicals, and there's no genuine sense in which the genes for testosterone and oxytocin could be said to be in conflict, but this cartoon sketch of Margulis vs. Dawkins still has a certain mischievous joy to its imagery.

The foregoing discussion goes some way towards demonstrating why co-operation is so well represented throughout biology: it is an advantage so powerful that whenever species stumble upon co-operative benefits they tend to endure. Rather than trying to shoehorn this observation into older models (as with kin selection), perhaps we might consider a new myth which captures the incredible role of co-operation in evolution. Simple forms of life co-operate because it's in their mutual benefit and advantages tend to persist, but more complex creatures (such as humans and other mammals) co-operate not just because it is immediately advantageous but also because we have inherited a biological mechanism which facilitates the formation of trusting relationships – primarily with our relatives, but potentially with anyone. The story that *trust is an advantage* runs contrary to the usual mythologies of evolution, but it is just as valid, and far more optimistic.

The Secret of Life

In the transition from the gene's eye view to the world's eye view we have changed from seeing life from the bottom up, where the focus is the biochemistry that maintains the chain of inheritance, to viewing it top down, where the focus is the living planet itself, which the Gaia principle offers as a metaphorical mega-organism (Lovelock, 1979) or super-ecosystem (Margulis, 1998). The essence of gene supremacy lies in the degree of importance ascribed to molecular biology, particularly in the view that genetics is more important than what actually goes on in the lives of animals and other organisms. As already suggested, gene supremacists are generally critical of the mythology of Gaia, and indeed uncomfortable with any view of life that is not ultimately grounded in genetic explanations.

Bioethicist Joanna Zylinska (2009) has offered a compelling account of the historical circumstances by which gene supremacy forged its unique mythology, noting that up until the mid-twentieth century biology had not been taken very seriously by scientists, who generally viewed it as 'softer' than chemistry and particularly physics, which had enjoyed such spectacular successes at the turn of the century in the form of Einstein's theories of relativity and the mathematical formulation of quantum mechanics. Biology needed a new mythology that could allow it to tap into the respect that 'harder' sciences had been enjoying. Physicist Evelyn Fox Keller (1992), from who's thought Zylinska's account draws heavily, puts the matter as follows:

To historians of science, the story of real interest might be said to lie in the redefinition of what a scientific biology meant; in the story of the transformation of biology from a science in which the language of mystery had a place not only legitimate but highly functional, to a different kind of science – a science more like physics, predicated on the conviction that the mysteries of life were there to be unraveled, a science that tolerated no secrets. In this retelling, the historian's focus inevitably shifts from the accomplishments of molecular biology to the representation of those accomplishments.

Zylinska notes that two diverse perspectives came together to allow for this transformation of the prevailing biological myths: the newly emerging conceptual entity 'the gene', and a new view of life offered by prominent quantum physicist Erwin Schrödinger (1948), who began to suggest that life could be thought about in terms of *information*, an idea emerging from early (pre-genetic) observations of mutation that Schrödinger thought to be "easily the most interesting that science has revealed in our day". Zylinska suggests that many physicists, feeling that physics had somehow been tainted in the wake of the atomic bomb, switched to studying biology – bringing with them "physics' authority and methodology".

The biologist James Watson, who was to play a key role in developing the new mythology, was greatly inspired by Schrödinger's approach, and was taken with the idea that "life might be perpetuated by means of an instruction book inscribed in a secret code" (Watson, 2004). At a time when information theory and cybernetics were rising stars in the scientific world, the reconceptualization of biology in terms of a molecular message written in a code that had to be cracked was appealing. The invention of computers during World War II as a means of deciphering coded communications had created a considerable interest in decryption that served as the backdrop for the new biological myths.

Once life could be thought about from an information perspective, the rich complexity of the life of organisms could be substituted for the far simpler laboratory problem of reading the message written in the DNA molecule, which had been singled out for attention following experiments carried out by group of researchers lead by Oswald Avery (1944) who identified DNA as something that bacteria were exchanging that led to new traits being expressed. Keller notes how the importance of DNA research became inflated by making the experimental research seem equivalent to "cracking the secret of life". Indeed, as Watson reports, when he and his collaborator Francis Crick discovered the chemistry of the base pairs that make up the double helix structure of DNA, Crick announced to everyone in *The Eagle* pub next to their Cambridge laboratory on February 28, 1953 that they had found "the secret of life".

Thus research in genetics became branded as 'the secret of life', and gene supremacy grew in influence. At root, this arose out of rhetoric intended to raise the profile of biological science and thus secure lucrative funding, but it was also (as Zylinska notes) "an attempt to capture public imagination through the daunting metaphysical connotations that 'cracking the secret of life' entailed." Indeed, this is why even though many of the scientists did not really believe that this was a fair description of the work that was being conducted, they continued to use the myth. The 'genetic code' metaphor borrowed some of the excitement generated by cryptography and information theory, and helped gene supremacy capture the public imagination. Even today, the prevailing manner in which Watson and Crick's work is presented in the media uses the mythic images of 'cracking the code' and 'the secret of life'.

It is important to reflect just how much is lost in this perspective. Gene supremacy elevates the importance of molecular biology above all other aspects of life – and even if we constrain our attention to scientific research, this effectively downplays the importance of the study of animals themselves (zoology) or their behavior (ethology), despite the fact that *animals are not reducible to their genes*. As Ruse (2003b) notes "organisms are package deals" – even though genes play important roles in the development and biology of animals, it is simply not possible to treat genes as a substitute for the creature itself. Even if we had a complete specification of all the genes in a particular animal, we would know very little about that animal. As Lewontin (1991) points out "even if I knew the genes of a developing organism and the complete sequence of its environments, I could not specify the organism".

Beyond the way that gene supremacy distorts the science of life, Zylinska suggests there may be another risk associated with the mythology of 'the secret of life'. By affording such weight to the decoding of genetics, gene supremacy has had political ramifications that are scarcely noted. What was once the "private realm of the flesh" has become open to a kind of imperial colonization in which real lives become secondary to "a calculable biological entity described as a 'population'". What is presented publicly as acts of scientific discovery can also be seen from an ethical perspective as an act of control – a perspective already apparent in the concept of sociobiology (E. O. Wilson, 1975), which purports to study the behavior of social groups using concepts from population genetics *as if behavior were reducible to genes*.

This hidden element of political power lurking in the background of genetic research was foreshadowed by Ivan Illich (1975) who recognized that changes in beliefs about health over the past few centuries have culminated in a medical regime that at times can be serious threat to the health it supposedly seeks to improve. Statistical data, being more manageable than actual people, becomes the venue in which medical performance is judged – the actual health concerns of individuals cease to be important once life expectancy becomes a commodity to be maximized. The danger in treating illness as primarily a matter for experts is that individuals cease to be responsible for their own health, resulting in a disempowerment that can exclude individuals from important decisions about their own lives and health. As Zylinska observes, talk of codes and secrets suggests a clear line separating "those who can decipher the code and those who can only be awed by it".

Perhaps the most famous example of gene supremacy distorting health issues is the Human Genome Project (HGP), which was completed in 2003 after thirteen years of research. A *genome* is simply the name for all the genes of an individual organism, and HGP's goal was the mapping of all the genes relevant to human biology. On the one hand, it was heralded as a major scientific breakthrough, and a landmark in our biological understanding; on the other, it was criticized on ethical and practical grounds, and accused of being radically less valuable than had been claimed.

I share with the geneticist Richard Lewontin a deep skepticism of the value of the Human Genome Project, which has always been woefully naïve in its assessment of what the $2.7 billion project could achieve. Lewontin (2000), as an expert in genetics, was never won over by the argument that the project was justified on the grounds of producing health benefits, seeing this as a paper thin justification created to validate what is essentially just a hugely expensive intellectual exercise. Part of the problem in this regard is that *everyone's gene patterns are different*, and there is no way of determining a 'standard' gene pattern to compare with. Without this, claims to reveal the genetic causes of various disorders are difficult to substantiate. Dismissing it as "a piece of direct-mail advertising" Lewontin declared:

It is simply impossible to justify the expenditure of a trillion dollars on a project to put in sequence the complex DNA of a 'typical' human being or corn plant on the grounds that it would be a lovely thing to behold. So we are assured that it is really all in the interest of curing cancer, relieving schizophrenia, and making groceries cheaper...

This is the first and most obvious criticism of the HGP: it doesn't map *the* human genome, it maps *a* human genome. Much of the DNA for the public project came from an anonymous male donor from Buffalo, New York, codenamed RP11. One person jokingly noted in respect of the project that once all the chromosomes were fully mapped and sequenced "they'll tell us everything there is to know about one French farmer and a lady from Philadelphia." (quoted anonymously in King, 2002). We can't claim to have complete human genetic knowledge until we've mapped a reasonable sample size, and that isn't what's been done. In fact, because of the genetic variation between individuals, there simply can't be one definitive human sequence – and suggesting there can be creates serious questions about an appropriate definition for what should be considered 'normal' – a criticism at the center of Zylinska's ethical objections to the project.

This is just the tip of the iceberg when it comes to the issues, however. The HGP claims to map a complete genome, but in fact what it has focused upon is identifying the 20-25 thousand genes present in a particular human's DNA – and this isn't necessarily the same as the complete genome. In fact, it's about 5% of the genetic data. While some gene supremacists may think this is sufficient, this view rests on very simplistic ideas about DNA. The *non-coding sections*, conventionally dismissed as 'junk DNA', actually serve very important roles both in terms of the evolution of new genes (an issue we will return to) and also for the great many regulatory signals that are vital to understanding cellular biology.

To cast light upon some of the major misconceptions, I'd like to make an analogy between the data contained within DNA and a telephone directory. Mapping the genetic data is much like collating a list of telephone numbers – that volume of data is certainly useful, when you know what it's for. But having a human gene sequence without the biological knowledge to go with it is like publishing a telephone directory without inventing the telephone. Actually, if you examine some of the bizarrely optimistic claims concerning the possibilities of genetic engineering that might be unlocked by the Human Genome Project, the metaphor becomes something like the belief that you can work out how to build a telephone by studying the telephone directory.

Because medical benefits were not the motivation to map the genome, but rather the excuse for spending so much money doing so, those with an interest in healthcare have been particularly critical. Abby Littman of Montreal's McGill University is another outspoken critic of the project, contending that "everything that's been done so far is about managing the genome project instead of questioning the whole issue of whether there should be a human genome project." (Pellerin, 1994). If the goal is better health, Littman suggests the molecular geneticists involved have focused on the wrong part of the problem. She asks:

Why are we so busy mapping the genome? Why don't we map the environment instead of mapping the genome and worry about things that really make us sick that we don't know anything about? Why do we think it's so much easier to change genes than environmental conditions that put us at risk? Because it's more expensive to clean up the environment than to deal with people who are at medical risk because of the environment.

Her view is that most human disabilities happen after we are born, and are caused by accidents, injuries or environmental factors – not as a consequence of specific gene patterns. She also questions the eugenic implications of this kind of project: "Do we want to live in a society where nobody is born with Down's syndrome? If so, why? That's an ultimate aim of these tests. Does this make us a better society? I'd like to make geneticists think about these questions as they do their work."

If the critics are correct that the health benefits of the HGP to the public are far lower than has been suggested, who are the beneficiaries of this research? One obvious candidate is biotechnology and pharmaceutical companies. Lewontin (1991) notes that "no prominent molecular biologist of my acquaintance is without a financial stake in the biotechnology business." The data produced by this research might have been extremely valuable to pharmaceutical companies as a potential source of new drugs, and to other biotechnology companies, but even this value appears to be less than was anticipated. Craig Venter, the president of Celera, the company which conducted the private portion of the genome mapping, stated that "the drug industry has been saying 'one gene, one patent, one drug', but the uses for this approach can be counted on the fingers" (Parsons, 2001) and also that "the notion that one gene equals one disease, or that one gene produces one key protein, is flying out of the window." (Henderson, 2001).

More likely beneficiaries from the research are health insurance companies, who can potentially use information obtained from a sample of DNA to reject people who show genetic predisposition to certain medical conditions. Thomas Murray suggests that "the most important movement in the ethics of workplace genetic testing has been away from the original vision of a public health measure, to screening as a way of reducing illness-related costs with no effect on the overall incidence of disease." (Pellerin, 1994). So now we are looking at the Human Genome Project not as a great advance in health care, but as a means of insurance companies denying medical cover – the health benefits for the general public are thus *worse* than they were before the project began.

All this makes it sound that nothing good has come from the Human Genome Project, but there is perhaps a silver lining: by exposing the naïvety of overly simplistic interpretations of genetics, the HGP has helped improve our understanding of biology by undermining old fashioned deterministic myths. Venter notes, in assessing the impact of the project his company has assisted in completing:

There are two fallacies to be avoided: determinism, the idea that all characteristics of a person are hard-wired by the genome; and reductionism, that now the human genome is completely known it is just a matter of time before our understanding of gene functions and interactions will provide a complete causal description of human variability. In everyday language the talk is about a gene for this and a gene for that. We are now finding that that is rarely so. The number of genes that work in that way can almost be counted on your fingers, because we are just not hard-wired in that way. (Henderson, 2001).

We have learned that organic biochemistry is not as simple as dialing a telephone number. There are complex multi-dimensional relationships between the tens of thousands of genes involved in building and regulating our bodies, and these can only be fully understood when taken also in the context of the conditions we live in. To understand an organism requires more than just knowledge of its genes, we must know how those genes interact and how those interactions relate to the organism's environment. The assumption of the linear influence of individual genes has suffered a fatal, yet long overdue, blow. Critics of the Human Genome Project have been saying this for decades now, but the biotechnology community has simply ignored the practical – and ethical – implications of their work, in the blind lust for either knowledge, money, or both.

A statement issued by Celera in 2001 admitted "because of the relatively low number of genes... it will be necessary to look elsewhere for the mechanisms that generate the complexities inherent in human development." Responding to this, Ari Patrinos of the US Department of Energy (who funded much of the public research) said: "It's kind of humbling, isn't it?" (Parsons, 2001). Any project that can inspire a meek outlook among scientists is certainly exceptional, but at a cost of nearly three billion dollars there must surely be cheaper ways of teaching humility. The fact that many geneticists (including Lewontin) had always been doubtful of the value of the Human Genome Project is a sobering reminder of just how powerful the myth of DNA as 'the secret of life' has become.

Ultimately, what allowed the HGP to conduct its multi-billion dollar funding heist was the extent to which gene supremacy had been allowed to saturate the biological mythology. Katrin Weigmann (2004) is frank about the way scientific metaphors all too easily exceed their usefulness:

Genetic metaphors often convey the impression that there is much more potential or many more implications in genetics and genomics than is really so... The exaggeration of the potential of molecular genetics is achieved by overemphasizing the power of the gene and also that of the scientist analyzing it.

The very phrase 'genetic engineering', for which the HGP can be imagined as a rather expensive fact finding mission, is problematic. Midgley (2003) notes that it conjures up analogies with machines that are not appropriate, since the elements of machines – cogs and sprockets – are more or less fully understood by their designers. We are a long way from having the same degree of understanding about molecular or cellular biology. Furthermore, there is an ideology of control behind the image of bio-engineering that concerns Midgley and Zylinska alike. This way of looking at life "is not itself a piece of science but a powerful myth expressing a determination to put ourselves in a relation of control to the non-human world around us, to be in the driving seat at all costs rather than attending to that world and trying to understand how it works." (Midgley, 2003).

The historical relationship between information theory and the genetic 'code' presents a mythic image, one also suggested explicitly at times by Dawkins, that cells are a kind of hardware for which DNA serves as the software. Having deciphered the code, we are supposed to be able to write our own programs. Lewontin (2002) parodies this belief succinctly: "Any computer that did as poor a job of computation as an organism does from its genetic 'program' would immediately be thrown into the trash and its manufacturer would be sued by the purchaser."

The genetic engineering we are currently conducting is not really akin to what a programmer does in writing a functioning software application – it is much closer to what a hacker does when they subvert an existing app they didn't write, and don't necessarily understand. In part because what is conducted amounts to little more than experimental tinkering, there are serious ethical issues surrounding modern genetics that stretch far beyond just the Human Genome Project. David Heyd (1992) sternly proposes that: "tampering with the natural biological process of species evolution and individual evolution... [is] a form of metaphysical trespass," while the Senior Editor of *Science*, Barbara Jasny, admitted that "the Human Genome Project [stretches] the limits of the technology and the limits of our ability to ethically and rationally apply genetic information to our lives." (Pellerin, 1994).

Part of the problem, as Midgley (1985) recognizes, is that

supporters of genetic engineering place their faith in *techniques* without recognizing that these are means, not ends. Believing that bio-engineering is beneficial is one thing, but what are we going to use it *for*? As Midgley chides: "What we need is to hear about is aims... What we get is a recommendation to entrust change to a certain set of experts, whose training has not called on them to pay any attention to conflicting aims at all." Zylinska concurs, suggesting that when it comes to matters of life and health, they cannot be left just to experts – "all free thinking citizens in liberal democracies need to have a say about them" (Zylinska and Bateman, 2009).

Although his books have done much to foster the mythic image of genes, Dawkins is also concerned about the potential dangers of unfettered genetic engineering. He suggests the essential paradox of the new technology is that it is potentially "too good at what it does" and that the "formidable power" of the technique "makes formidable demands on society's wisdom." He stresses the importance of making the right choices about how the new technology might be used, and recognizes that there are significant political issues in determining *who* will make those decisions. Dawkins warns that if questions about bio-engineering are left to the marketplace, the environment is certain to suffer, adding that this "is true about so many aspects of life". (Dawkins, 1998).

The story of DNA as the 'secret of life' is not a myth of evolution, as such, but it is closely related to the myth of gene supremacy that emerged largely from evolutionary studies. In fact, both were symptoms of the same fervor – that which sought to find a solid, chemical basis for the inheritance mechanisms that had been assigned to the metaphorical 'gene' long before Watson and Crick figured out the chemistry of the base pairs within the double helix. When the laboratory returned a triumphant answer to the question of what allowed the chain of inheritance to function, a kind of gold rush began on molecular biology. Only now is it becoming clear that genetics, while certainly a valuable addition to human knowledge, is far from a universal remedy.

Flexible Mutants

The enthusiasm about DNA in the twentieth century severely distorted the way biologists thought about the nature of life. In some respects, the selfish gene myth was the epitome of this confusion, since it implied two wildly mistaken assumptions: that all life is fundamentally selfish (the myth of egoism), and that the behavior of creatures is best explained by reference to the selective value of their genes (the myth of gene supremacy). Egoism, while deeply metaphysical, is not a biological theory, so in terms of the scientific content of the selfish gene it is gene supremacy that requires the most scrutiny. There is, I will reiterate, an important scientific perspective at the heart of the selfish gene, namely the gene's eye view – but there are problems even with the science here that need to be taken into account.

Recall that the reason Dawkins choose 'the selfish gene' as a means of expressing George C. Williams intuitions concerning the centrality of the gene to evolution was that he felt it captured the idea central to inclusive fitness: the evolution of an organism can be expected to occur in such a way as to maximize the number of copies of its gene variants in future generations. It is in this metaphorical sense that Dawkins uses 'selfish' – from the gene's eye view, the more closely related two individual animals are at the genetic level, the more reasonable it is for them to behave selflessly towards each other. This leads to the myth of kin selection, namely, that animals co-operate solely with their genetic relatives

Dawkins himself does not fall prey of this myth, being well aware that co-operation is an important strategy, even when it is not with genetic relatives – his documentary *Nice Guys Finish First* (Horizon, 1986) argues expressly that evolution often favors co-operation. He has also admitted that the title of his book, *The Selfish Gene*, "might give an inadequate impression of its contents", and has suggested that in retrospect he should have taken Tom Maschler's advice and used the title *The Immortal Gene* instead (Dawkins, 2006). The intent, after all, was to discuss the circumstances by which certain genes are propagated, and the metaphor of a selfish gene is inadequate to this task, being fundamentally misleading.

If 'selfish' isn't an effective metaphor for genes, what is? To answer this question, it is useful to examine what actually goes on at the genetic level of life. One of the most surprising realizations to have emerged from the research community's obsession with cataloguing DNA has been the astonishing extent to which genes are *conserved* across different species. Conservation, in the context of genes, means long-term persistence, and genes that are 'highly conserved' appear in the genetic structure of many different species. One particular set of genes that have been highly conserved over the incredible span of time since life appeared on our planet really drives home the extent to which all life is related.

At the end of the 1970s, research biologists Christiane Nüsslein-Volhard and Eric Wieschaus were conducting a series of genetic experiments on fruit flies. Flies, being short lived animals, are popular for lab research on evolution since multiple generations can be produced in a comparatively short period of time. Nüsslein-Volhard and Wieschaus discovered a set of fifteen gene positions in the DNA of fruit flies that when mutated altered the way their larva segmented (Nüsslein-Volhard and Wieschaus, 1980). Their contemporary, Edward Lewis, had more or less simultaneously discovered genes that govern the transformation of larval segments into the body parts of the adult fruit flies (Lewis, 1978). The three biologists won the Nobel Prize in Physiology or Medicine in 1995, having discovered what are now called *hox genes* – a group of related genes that determine the body plans of all multi-cellular organisms.

These discoveries led to the astonishing discovery of what has been called the developmental-genetic toolkit, a set of genes shared (in one form or another) with every living thing to have emerged since the Cambrian explosion. This *genetic toolbox* revealed something that no-one in biology had even remotely suspected – common elements like eyes and legs that occur in wildly different animals weren't just similar features that evolved independently, *they were the result of a common ancestry.* Despite the eyes themselves being wildly different in form, the compound eyes of fruit flies and the camera-like eye of mammals shared remarkably similar DNA. As cell biologist Stuart Newman observes, a gene variant that causes flies to lose their eyes or form smaller eyes has "extensive DNA sequence similarity" to genes that reduce the size of eyes in mice or humans (Newman, 2006).

Newman even claims that the genetic toolbox gives us an explanation for the Cambrian explosion, and of large-scale evolutionary changes in general. Rather than the profuse diversity of life that appeared around 500 million years ago being the result of the kind of mechanisms considered in the gene-centered view, he suggests it is more plausible to imagine an ancestral form in which the toolbox genes initially served some unrelated purpose. In the early days of multi-cellular life, it is possible that a single genetic pattern could have produced a variety of different body plans, simply as a consequence of the material properties of the proteins in the ancestor's cells and the conditions of the environment. If this is correct, then "large-scale evolution could have occurred relatively rapidly, with only minimal change in the molecular components of the underlying developmental mechanisms". What's more, this perspective suggests major transitions in the history of life can be abrupt and even influenced directly by changes in the environment – irrespective of the usual selection of genetic variants (Newman, 2006).

Understandably, astonishing claims such as these meant that the genetic toolbox sparked a significant conflict among biologists studying evolution, an argument which Ruse has characterized as "form versus function" (Ruse, 2003b). In fact, as Ruse notes, this is a rather old discussion, one that dates back to the early days of evolutionary biology but that has now reappeared in the light of toolbox genes. Advocates for function maintain the interpretation of evolution in terms of genes and adaptation critiqued in this chapter, and explored from a different perspective in the next. Advocates for form like Newman contend that gene supremacy has led to "an impoverished evolutionary theory" and that animals "are in no fashion reducible to genes" (Newman, 2009).

However, as with so many bitter fights among scientists, the disagreements are not as substantial as it may first appear. Certainly, this is not a fight about whether or not natural selection occurs – despite Darwin's metaphor of fitness to environment offering little more than an informative fiction, no evolutionist disputes that it happens or that it is important. What is in dispute, however, is whether selection has been afforded too much attention, and whether a fully developed theory of evolution would place natural selection in a central role. Ruse, who largely sides with the gene supremacists in terms of preferring to interpret evolution on the model of individual selection, admits that the new molecular biology of development is "incredibly exciting", but states that natural selection still has the key role – it "winnows out" the unsuccessful from the successful. He suggests that "natural selection has been invigorated and made even healthier precisely because of the new emphasis on form," a position sure to frustrate opponents of gene supremacy.

The reason Ruse is able to offer such a nonplussed response to what might seem at first glance to be a significant upsetting of the apple cart is that the myth of gene supremacy was always in the background: the foreground was always natural selection, and specifically the way that natural selection leads to adaptations. The genetic toolbox argues against some of the tacit assumptions of gene supremacy, but it doesn't argue against selection as such. And since the toolbox itself is comprised of genes, gene supremacists can rest secure in the belief that DNA *really was* the missing part of Darwin's theory. What has changed – almost unnoticed – in the last fifty years is that we have completely redefined what *genetic determinism* is assumed to mean.

In the early twentieth century, H. J. Muller published a paper entitled 'The Gene as the Basis of Life', which expressed the view that *something* in the protoplasm of the cell constituted genes, and that these putative genes held "the primary secrets common to all life" (Muller, 1926). At this point – prior to the discovery of DNA – there was already a detectible perspective in which genes (whatever they might turn out to be) were offered as the central explanatory principle in biology. This is the 'classical' view of genetic determinism, one that is found fifty years later in Dawkins' *The Selfish Gene*, epitomized by quotes such as "we are survival machines, robot vehicles blindly programmed to preserve the selfish molecules known as genes" At this time, confidence in the explanatory power of genes is at its peak – hence sociobiology and its successor evolutionary psychology, that purport to explain the behavior of humans and other animals in terms of the gene's eye view. This time – the mid-seventies – was the heyday of gene supremacy.

Since then, however, skepticism as to how much the gene alone can contribute to our understanding of the nature of life has grown. As already noted, it has become clearer and clearer that the gene isn't enough by itself to fully explicate any animal. As a consequence genetic determinism has changed its tone: rather than *determining* behavior, talk has changed to genes *influencing* behavior, rather than single genes being assumed to explain single traits, traits are assumed to occur as a confluence of many different genes working in concert. This new perspective is more restrained than the fervor of the seventies, but gene supremacy survives in its new, tempered form, which rests on the dependency of evolution upon the gene. As Ruse puts it: "no gene change, no evolution", that is, whatever else you say about the gene, you can't get around its relevance to natural selection.

There is, however, one more thing to take into account before we leave myths of gene supremacy behind. Up to this point, we have assumed that genes code for proteins the way that, say, a telephone number specifies a single caller. If we change just one number in the sequence we dial, we get through to someone completely different. Similarly, the assumption up to this point has been that specific genes code for specific proteins – if you want to produce (say) adrenalin, you need a particular "address" in the DNA that encodes that chemical. (Actually, adrenalin is synthesized in the adrenal gland by a series of reactions acting on other chemicals, but there is still a raw material with its own gene at the start of this process, so the analogy is close enough for our purposes). It is tempting to think that a change or *mutation* to the gene for adrenalin would produce a different hormone, but it transpires that the relationship between genetics and proteins is not that simple.

In 1968, the Japanese biologist Motō Kimura suggested that the vast majority of the molecular differences in the genetic code of existing species of animals were selectively 'neutral' – that is, they had no significant effects either at the level of biochemistry, or at the level of the organism itself. Kimura reached this conclusion from examining the way DNA codes for amino acids, the building blocks out of which proteins are 'built', and noticing that different sequences in the DNA could produce *the same* amino acid. For instance, the most common amino acid found in proteins, known as leucine, has six different genetic 'recipes' (which can be written as UUA, UUG, CUU, CUC, CUA, CUG). As a consequence, unlike telephone numbers, it is possible to make changes to the genetic code *without* changing the principle effects of the resulting protein. This is known as the neutral theory of molecular evolution, or just the *neutral theory*.

As with any change in perspective among scientists, the neutral theory provoked heated debate. The argument wasn't about whether natural selection occurred, but about how *frequently* it occurred, and about the significance of what could be termed, following Kimura, *neutral mutations*. One camp, the selectionists, remained adamant that natural selection was the primary or only cause of evolution, while the other, the neutralists, maintained that neutral mutations were widespread, and that random sampling of different gene variants – what is called *genetic drift* – was important to the evolution of specific proteins.

The heart of Kimura's argument was that much of the variation in DNA is never subject to selection, and hence varies randomly, a claim that was validated by looking at non-coding sections of DNA. These parts of an organism's DNA that don't specify or regulate the production of proteins change at a far more rapid rate than the parts of DNA that are called genes. It is easy to see why this should be so: if a mutation to an important gene stops it from functioning, the creature affected by that mutation has to make do without the protein or function in question. Because so much of animal's biology is rather important to its survival, most of these deleterious mutations are *fatal*, and the mutant gene variant never gets passed on. Conversely, non-coding sections don't have any direct effect on organisms and are free to change.

The selectionists, rather dismissively, coined the term 'evolutionary noise' to describe neutral gene variants, and Kimura (1983) noted that many biologists told him that the neutral theory was not biologically important, since it was (by definition) nothing to do with adaptation. He rejected this claim, stating that it was "too narrow a view" and maintaining that since what is important in science is uncovering the truth, the value of the neutral theory should rest on its validity as a hypothesis, and nothing else. (This idea of adaptation as being the *important* aspect of biology is the theme of the next chapter).

It is worth noting, as Peter Godfrey-Smith (2001) has observed, that Dawkins "sees himself as having nothing invested in this debate". Dawkins is the archetypal gene supremacist, but Kimura's denial of selectionism isn't problematic for Dawkins since the neutral theory amounts to a statement concerning genetic variation considered globally and inclusively, across all cases of molecular change. This doesn't bear on gene supremacy at all since even if selection "might only explain 1% of all molecular genetic change", as far as gene supremacists are concerned "this is the 1% that counts." It may seem, therefore, that the neutral theory is an irrelevant sideline, but there is a point to this digression.

In 2008, Andreas Wagner proposed a reconciliation between neutralism and selectionism. Wagner's proposal was based on work he had conducted using sequences of RNA, the single-stranded cousin of DNA that folds into interesting shapes precisely because it doesn't have the neat double helix form. By using a computer algorithm that predicts the shape of RNA sequences, and models of molecular variation in the RNA that are functionally equivalent to genetic variation in DNA, Wagner was able to explore the effects of mutations on the biochemistry of living systems. His results showed that some of the folded shapes that the molecules adopted were significantly more robust to mutation than others, which meant that – like amino acids having multiple 'recipes' – certain biochemical functions could be represented by many different genetic sequences.

Rather than selectionism versus neutralism, Wagner proposes that evolutionary change may proceed by successive sequences of "neutralist regimes" and "selectionist regimes". During a "neutralist regime", neutral mutations accumulate without any of the important gene functions being disrupted. As a result, the diversity of gene variants increases significantly even though – from the perspective of the animals in question – nothing significant changes; the biology and biochemistry remains the same because although the genetic 'recipes' have changed, the 'dishes' remain the same. When the increased genetic diversity eventually leads to some significant advantage – or, in the fitness metaphor, when a gene variant of higher fitness results from a mutation – a "selectionist regime" occurs, and in a relatively short space of time (geologically speaking) the new advantage spreads throughout the population.

According to this new perspective, "neutrality facilitates evolutionary innovation and adaptation" because "robust molecules tolerating many neutral mutations more readily evolve new functions both on laboratory and evolutionary timescales." This helps resolve a long standing problem about how beneficial adaptations can actually occur in nature. The standard explanation, eloquently described by Dawkins in *Climbing Mount Improbable* (1996), presumes a series of simple steps, each of which is beneficial, such that these gradually accumulate into more advanced adaptations. Dawkins uses the example of an eye, showing how various different degrees of complexity can still provide an advantage to an animal, from light sensitive patches through pinhole cameras to lenses and focus. However, under selectionist thinking every change *must* be beneficial or it will not last, so gradual steps must be found for every evolving feature, and in many cases this explanation asks for a leap of faith.

Wagner's alternative framework eases some of the pressure, since changes don't have to provide an advantage at every step along the path. Neutral mutations build possibilities that are eventually cashed out in terms of beneficial adaptations, since large quantities of neutral (or nearly neutral) mutations can accumulate over some time before one additional change makes all the difference. It's like playing gin rummy with an unlimited hand size: it's much easier to form sets or runs the more cards you have. The acquisition of neutral mutations during a neutralist regime is like drawing extra cards, increasing the hand size, while the selectionist regime is like trimming the deadwood back so you can go out with the best possible hand. Wagner's *theory of genetic innovation* suggests it's not so improbable that beneficial mutations will eventually occur – if you draw long enough from a big enough deck, you eventually go out with a hand that's all aces.

But there's a catch – Wagner's theory makes adaptations more plausible, but it seriously undercuts some of the core principles that the gene-centered view depends upon. As Williams (1966) put it, the gene is the only candidate for selection because it is a "necessary condition" that whatever it is being selected "must have a high degree of permanence". Individual combinations of genes within specific animals are utterly unique; only the patterns of the individual genes endure long enough for natural selection to work on them. This durability of the gene as an information pattern is why Dawkins now thinks *The Immortal Gene* would have been a better title than *The Selfish Gene*. Yet Wagner's theory shows that *even genes are subject to significant changes*, since neutral mutations alter specific genes in ways that don't affect their usefulness to the animal in question (their 'fitness') but *do* alter their underlying genetic 'recipe'.

It follows, according to Wagner's theory of genetic innovation, that genes just don't have the high degree of permanence Williams had relied upon. Despite Dawkins' claim to their immortality, genes are mortal after all – they change over time. What is preserved are the functional consequences of specific networks of genes, because it is these that provide biological advantages for the animals carrying the genes – as can be seen with the toolbox genes, which have vastly different variants across all the multi-cellular animals but maintain a constant function, that of specifying body plans. That function is conserved; the genes that provide it are subject to change. As Newman observes, it is frequently the case that "mutation or even deletion of a gene leads to little or no detectable change" (Newman, 2009).

Wagner offers a metaphor to express his model, suggesting that the space of all possible gene patterns is "partitioned" into multiple networks of genes that correspond to particular actual functions. Beneficial mutations occur as a result of a collection of (initially neutral) mutations that eventually 'nudge' one gene network into an entirely new network with a *different* function that is then subject to natural selection. In terms of the selection metaphor, the new network has higher fitness and so becomes more widespread because, as my myth has it, advantages persist. However, once the new beneficial function is in place the genes that specify it are once again subject to change by neutral mutation, and the evolutionary gin rummy game is on again.

Earlier, I asked: if 'selfish' isn't an effective metaphor for genes, then what is? Looking at the changeability suggested by Wagner's theory of genetic innovation, and the astonishing variety of life that results from the genetic toolkit, from cats and dogs to hamsters and swifts, I would have to suggest that the trait that most stands out is the flexibility surrounding the biological functions that collections of genes make possible. Perhaps instead of the myth of the selfish gene, we should think instead of the flexible gene.

4. Metaphors of Design

Intelligent Design

The wing of the swift resembles the wing of a fighter jet not by coincidence but because it is an effective shape for something intending to fly fast and make agile turns. Other birds, those which do not race around the sky with the speed of the swift, have different shaped wings, each reflecting different needs. The elliptical wing of the crow is suited for tight maneuvering in confined spaces, such as among trees; the extra-long wing of the albatross is perfect for gliding but requires a long taxi to get airborne, and are thus more easily used over the ocean; the eagle's wide, slotted wings are perfect for soaring and a quick take-off. Airplanes often have broadly similar shapes, because the aerodynamics of flight are the same for machines as they are for living creatures.

This explains why planes and birds have resemblance, but it does not explain *why* birds have wings that are so perfectly suited to this kind of flying. The evolutionary explanation of *adaptation* says that the wings conform to the environmental circumstances within which the bird can be found, and that this happens as a consequence of natural selection. Those birds that were better suited to the relevant circumstances thrived and thus, by differential survivorship, the wings gradually evolved to fit those conditions of life. This is Darwin's big idea in a nutshell.

However, this account has some limitations. For a start, Darwin's fitness to environment, which is invoked in explaining adaptation, is still only a metaphor. It doesn't actually answer the question of *why* the swifts are adapted for high-speed flight so much as it provides a way of imagining *how* it may have happened. A complete explanation requires knowing something about how the genetic toolbox is able to create the different kind of wings, an account of the circumstances that caused the ancestors of swifts to pursue this lifestyle, and more besides. Some of this information is attainable, and some of it can only be speculated. Often, talk of adaptation occurs as the pure speculation I have called How-Why games, and falls considerably short of fully explicating a particular feature (although this is not to deny that it might be possible to do so in certain cases).

Some people, however, claim the Darwinian account of adaptation isn't just limited, it is entirely misleading. They claim that there is detectable evidence of intentional design in the living world – that the swift is better explained by inferring a designer than it is by adaptation, natural selection and so forth. They claim unguided and unintelligent causes are insufficient to explain the wings of the swift and many other aspects of both the living world and the universe. Based on the evidence of design that can readily be seen in nature – like the swift's wing, which seems designed for high speed flight – they believe the best explanation must be some sort of intelligence, whatever it might turn out to be. This is the fifth myth of evolution, *intelligent design*.

It has become difficult for anyone with any connection to either science or religion to remain neutral in what has become one of the greatest storm-in-a-teacup controversies of recent decades. This issue is not as simple as is sometimes suggested, in part because partisans on both sides seem to have lost sight of what they are fighting for – if indeed, they were ever fully cognizant of their motivations. But this is not to say that it is too complex to be resolved, it just requires decomposition into the relevant issues. Once suitably deconstructed, it will hopefully become clear that there are criticisms to be leveled against either side of the divide this issue has fostered.

Before proceeding, let me be clear than when I talk about the Intelligent Design movement (with capital letters), I am referring to those people whose party line is that evolution is false and creationism is true, or who mount the same argument in softer lines – perhaps suggesting that evolution is one theory, and creationism is an alternative theory. It is not my intent to lump into this bracket those people of faith who are happy to talk of 'the glory of God's creation', or those who see evolution as the method that God used to put life into motion, or any other moderate position on these issues. When I talk of Intelligent Design (or ID), I mean the political movement, primarily in the United States, that rallies under that particular banner for the express purpose of opposing the teaching of evolution.

If you are the kind of person who gets a hot flush at the mere mention of intelligent design, you need to take a couple of deep breaths and set aside your preconceptions of what this topic is about. It is true that intelligent design is a flawed myth of evolution, and one that is in need of a more suitable alternative – in this respect, you can feel vindicated by your ire at the ID crowd. But if we take the time to understand the issues behind the fireworks, we will eventually discover that pulling up this problem by the roots will mean a lot more than simply trying to silence advocates of intelligent design – it will actually require that we attend to the original issue that motivated this movement, and that means addressing problematic beliefs that can be found on *both sides* of this debate.

Conversely, if you are the kind of person who believes that the study of the history and nature of life has been severely distorted by advocates of natural selection, you too need to take a couple of deep breaths and remember why this topic matters to you. If you are a person of faith and feel that 'neo-Darwinists' have distorted the facts, you can feel vindicated that there has been significant distortion of this kind. But you also need to remember that it makes no sense for a Christian or Muslim who has faith in God to claim that there is scientific evidence for God, especially a tenuous kind of evidence that still requires a leap of faith. Doubt is a normal part of a life of faith, but evidence is not:

what sustains the faith of a believer should be their experience of God in their lives, not forensic evidence.

The political aspect of the intelligent design furor no longer has much connection with the issues at the heart of the philosophy at question, and I will defer addressing the question of the purported war between 'science' and 'religion' until the next chapter, where it can get a proper hearing. Our concern for the time being is not the fight over the United States education system, nor the metaphysical commitments that drive that conflict, but rather the issue of the relationship between evolutionary science and the concept of design. And this particular topic has a history almost as long as civilization itself – both Plato (360 BC) and Aristotle (350 BC) debated the questions at the core of intelligent design, for instance.

For simplicity, I will consider the issue from which this topic springs the *design argument*, although it has many other names, most notably the teleological argument. We've already met this term 'teleology' in the context of How-Why games, which can also be called 'teleological games' since they deal with an implied purpose. The design argument is the ultimate How-Why game – rather than dealing in individual cases, why polar bears are white, or why lemons are sour, it attempts to tackle *every* case together, and draw conclusions about the apparent presence of purpose in nature. In one form or another, the design argument can be found in a great many of the religious traditions of the world. Christianity and Islam inherit it from the Greek philosophers, who were a significant (although often unrecognized) influence on the development of the religious cultures of the West.

There are distinct stages to the design argument: firstly, recognition that there is an apparent order or complexity to the world in need of explanation (the premise), secondly, the consequent deduction that the complexity that we see is evidence of design (the inference) and hence, by definition, a designer (the conclusion). Michael Ruse (2003b) has criticized the traditional names for the stages of the argument as being confusing and unclear, with compelling arguments, but his alternative solution involves changing the meaning of well-established terms, which is likely to make the situation worse rather than better. Anyone particularly interested in the history of the design argument is strongly recommended to read Ruse's *Darwin and Design*, since there is no better guide available, but for our purposes we shall treat the design argument as having three steps: the premise (the recognition of complexity), the inference (complexity implies design) and the conclusion (design implies a designer).

The Scottish philosopher David Hume criticized conventional Christian beliefs, including the design argument, in his posthumous work, *Dialogues Concerning Natural Religion* (1779). This book, despite the way it is sometimes presented, does not argue for atheism – Hume was a committed skeptic, and believed that outright non-belief required as much of a leap of faith (or perhaps, unfaith) as Christianity. It does, however, argue against conventional Christian views on God via a discussion between a trio of theologians each presenting a different view. Demea argues for traditional Christian orthodoxy, and receives a frosty reception – he storms out before the discussion is concluded. Cleanthes presents the design argument in its typical form. He is challenged in turn by Philo (who probably represents Hume's own views), who systematically attacks Cleanthes assumptions.

Many of the objections to the design argument that Philo offers remain relevant to contemporary discussions. In terms of the recognition of complexity, Philo observes that although we see order here on our world, it may be chaotic elsewhere in the universe, and besides, an inductive argument of this kind requires repeated experiences but the creation of the universe was a one-of-a-kind event (recall the sock drawer thought experiment). In terms of the implication of design, he questions the analogy with machines, and reflects that when we look at a ship we might be inclined to believe it the work of an ingenious carpenter – but then we would be disappointed to discover that he had simply copied an existing design, one that had been "gradually improving" after a long series of mistakes and corrections. In terms of the implication of a designer, he suggests that for all we know the apparent design in nature might be explicable by some principle of order within matter itself. Furthermore, even if the universe were of divine origin, that wouldn't allow us to infer that this cause was a single, all-powerful or entirely good being; indeed, the presence of pain and suffering in the world would appear to argue against this (what is known as the theological 'problem of evil').

As Ruse notes, Hume was "too successful" in mounting his argument, and he himself seems to have been aware of this. The problem is, all of Philo's arguments aside, there is *something* about the world – and particularly the living world – that did seem to point to an unexplained mystery. Hume had thrown down the gauntlet, but rather than demolish the design argument, he effectively renovated it, inviting fresh voices to tackle the problem anew. Two responses to Hume's critique are particularly important for the discussion of intelligent design. Firstly, William Paley, whose famous watchmaker argument was a major inspiration for Charles Darwin's work, and secondly the German philosopher Immanuel Kant, who we will consider shortly.

Paley was perhaps the most renowned exponent of what is termed *natural theology*, the process of reasoning from observations of the world to claims about God or gods. Paley, a Christian clergyman and philosopher, died just three years after publishing his most famous work, *Natural Theology* (1802), a book that in many respects is still the touchstone for contemporary arguments in favor of intelligent design. Paley observes, in the opening of the book, that if we were to find a watch, we would have no doubt from the intricacy of its mechanism that somewhere there was a watchmaker. So too, if we examine the complexity of the living world (Paley claims) we cannot help but conclude that somewhere there must be a world-maker. This *watchmaker analogy* is a metaphor intended to support the design argument.

Paley's watchmaker argument is precisely the claim of intelligent design: life shows evidence of having been designed, and from this we can reason to a designer. However, the contemporary Intelligent Design movement mounts its claims on interpretations of molecular biology and the like, and in this respect is far more advanced than Paley's theology. It also stops short of Paley's conclusion that this designer is God, although to my knowledge no supporter of ID has advocated an extra terrestrial or alternative identity, and critics tend to argue that the only reason to mount such an argument is to defend some kind of natural theology in the style of Paley. However, rather than examine the contemporary debate, which is complicated by strongly divergent metaphysical commitments on both sides, we may have more to gain by considering the arguments of Paley's contemporary, Immanuel Kant.

Kant provided the most thorough treatment of intelligent design prior to the last century in the second part of his *Critique of Judgment* (1790). The main focus of this book is aesthetics, and indeed the reason I came to read it was precisely because I had developed an interest in philosophy of art. However, I was really quite taken aback by the degree that Kant's philosophy is directly relevant to the contemporary debate on evolution. Kant does not speak of 'intelligent design', but rather of 'teleological judgment'. Again, 'teleological' refers to the study of purpose (and hence design), and especially the relationship between the appearance of design in nature and whatever can lay claim to being the ultimate cause of everything. The intersection with intelligent design is hopefully apparent.

In Kant's time, discussion of teleology was principally an ivory tower concern, but this is not to suggest that there are not parallels between the modern discussion and its eighteenth century forerunner, and this at many levels. For a start, the matters at the heart of the disputed claims have not changed significantly. Kant notes that if we do not attribute the cause of a particular natural entity being a certain way to a "teleological ground" then "its causality would have to be represented as blind mechanism." This observation foreshadows the conflicting camps in the modern intelligent design debate.

Beyond this, however, Kant has issues which mirror modern concerns in this respect. He notes that the basis of "our great admiration of nature" lies in part in the way we look at nature and easily conclude that it is "constituted just as if it were designedly intended for our use". However, he notes that this can go too far: "It is surely excusable that this admiration should through misunderstanding gradually rise to the height of fanaticism." The term *fanaticism* here means something very specific in the religious context Kant was writing from.

As Charles Taylor (2007) explains, the theology of the late eighteenth century identified three kinds of "dangerous religion": *superstition* (practices based upon faith in magic), *enthusiasm* (the certainty that you had heard the voice of God) and *fanaticism* (the kind of religious certainty that licensed going beyond the common moral order). In connecting the association of purpose in nature with fanaticism, Kant is expressly disavowing the theological justifications which lie at the heart of intelligent design as going too far. Yet Kant was a devout Christian, he certainly believed in God – he was simply not willing to endorse anything that goes beyond whatever reason can lay reasonable claim. To do so was, to Kant's mind, quite unwarranted.

Why does Kant not feel the need to draw on the purposiveness of nature as proof of God? He recognizes that this sense of purpose is there, after all, and he believes in God. What he does not agree with is the *method* by which such an assumption might be reached. In examining nature, he notes that much of what appears as purposive in nature could simply be the relationship between one element of nature and another. He notes that contemplation of nature can't allow us to conclude with certainty that there is some ultimate principle behind nature's actions, even if it "hypothetically gives indications of natural purposes". Considerably more would be needed to justify such a conclusion.

Kant's Restraining Order

This kind of shrewd caution concerning metaphysics is a hallmark of Kant's philosophy. Goaded into philosophical thought by the challenges Hume had laid down, Kant was keen to distinguish between the phenomenal world that we experience with our mind and senses, and what he called the 'noumenal world' – the real, material world 'out there' but strictly unknowable except via the phenomena we experience. Thus, claims about the material world that were not justified by phenomenal evidence were metaphysics, and had to be handled carefully. He had a definite place for such thoughts, but it wasn't in empirical science.

Before proceeding, we will have to consider some of his rather unique terminology since Kant, like rather too many philosophers, relies on some very carefully defined concepts to make his arguments. Kant uses the term 'natural purpose' to refer anything that could be construed to be both *cause and effect of itself*. He notes: "In such a product of nature every part not only exists *by means of* the other parts, but is thought as existing *for the sake of* the others and the whole, that is as an (organic) instrument." He further states: "Only a product of such a kind can be called a *natural purpose*, and this because it is an *organized* and *self-organizing being*."

This concept behind a "self-organized being" has a history that pre-dates Kant to a great extent, but only in terms of the related themes. It was Kant who singled out self-organization as a factor, and this appears to have been an influence on Ross Ashby (1956), who introduced it into modern cybernetics (which now uses the concept extensively). The comparison with design breaks down in respect of such self-organized beings – Kant notes that watches *are not like this*: "a watch wheel does not produce other wheels, still less does one watch produce other watches". An organized being was thus "not a mere machine", but something possessing the power to self-propagate its form by some means, to organize itself through some internal principle.

In this concept was the beginnings of the modern concept of an *organism*, and Kant has surprisingly nuanced views on the subject. On the one hand, he asserts that they are natural purposes (or natural ends) because we can conceive of the possibility of the existence of such things *only in accordance with the idea that they were produced by design*, but he denies that they should be thought of as being the products of conscious design at all, insisting they must be seen as products of nature. Kant essentially says that we have to think about animals *as if* they were designed, yet denies that we can conclude that they actually *were* designed.

Some might object to the idea that we *must* conceive of organisms in respect of design, but the evidence abounds in the language used to talk about the relationship between organisms and evolution. When we talk of a tail having the purpose of a counterbalance in mammals, or of incisors being used to cut food, or of camouflage as serving to conceal *we are talking in terms of design*. It may be the case that most modern folk attribute the *origin* of these features to adaptation and natural selection, but this does not change the fact that discussion of the features of organisms is universally couched in design terms. The rules of the How-Why games may have changed, since they are now more commonly played with evolution instead of God, but we still talk about the features of organisms by *thinking* in terms of design.

Note Kant's reluctance to proceed from the observation that we must talk about organisms in terms of design to the conclusion that organisms must have been consciously designed. He astutely recognized that the question of whether they were designed or otherwise was an untestable issue, observing "that would be to meddle in an extraneous business, in Metaphysic". Instead he states that it is enough to recognize that there are some objects which are "alone *explicable* according to natural laws which we can only think by means of the [design]".

Ruse observes that in recognizing the inescapability of design-thinking when it comes to organisms, "Kant got it right". It is because adaptations are explicitly "artifact-like" that we have recourse to a metaphor based on the idea of design: an artifact, like a watch, is designed for a purpose and so analogously (and metaphorically), we can talk about purposes when it comes to adaptations. As Ruse notes: "If the heart were not like a (human-made) pump, teleological language would never arise, but it is and so it does. What's more this kind of language is effectively unavoidable since, as the paleontologist Martin Rudwick (1964) stated in the context of fossils, it is "fundamental and unavoidable" that inferences about the functions or significance of various anatomical features will take place in terms of *what* they are capable of doing.

The *metaphor of design* is the myth I propose for replacing intelligent design. Physicalist objectors may complain that biology is categorically not forced to discuss purpose or to think in terms of design, since this kind of language can always be restated without reference to design or purpose. Yet as Ruse suggests, we can't throw out the functional language without ending up with an "emaciated analysis". Furthermore, Ernest Nagel (1961) demonstrates it is not tenable to suggest that every purpose-based description has a logically equivalent statement with no recourse to teleology, since if we attempt to restate physical laws in terms of design or purpose what we get is patent nonsense. Boyle's law sounds very strange indeed if we say "every gas at constant temperature under a variable pressure alters its volume in order to keep the product of the pressure and volume constant". There is something about purpose-focused thinking in biology that is indispensably connected with the study of life, as Kant asserts.

John O. Reiss (2005) raises a point of concern, however. Considering Nagel's teleological version of Boyle's law, Reiss recognizes that the reason this kind of purpose-focused explanation sounds odd here is that the behavior of gasses can be understood more-or-less deterministically, as the result of specific physical laws. These kinds of explanations involve an appeal to absolute necessity, and Reiss suggests that it is not appropriate to talk of purposes in these cases. What we require to justify talk of purposes is some kind of *conditional* necessity, since when we deal with deterministic systems "there is nothing left" for talk of purposes to help explain "that has not already been explained by the boundary conditions and the relevant laws."

The problem that Reiss draws attention to here is that we *only* need to invoke purpose when there is something left to explain beyond the consequences of mechanical, deterministic laws. We will have to look at Reiss' concerns in more detail later; for now, the metaphor of design that Ruse discusses isn't greatly troubled by this objection since in the case of explaining biological features we *don't* have access to a fully deterministic account of any kind. There *are* conditional issues to be explained in terms of how a particularly animal's biology ended up in a particular state, and as such we are at least provisionally justified in using teleological thinking of some kind in our attempts to understand biological origins.

Ruse's metaphor of design broadly agrees with Kant's view, which amounts to the claim that talk of design in life sciences is unavoidable, since biology can't reasonably be conducted otherwise. As Ruse suggests, thinking in terms of purpose and design is a necessity since those who study the life of animals "are as bound to teleology as they are to physics." For Kant, this doesn't mean giving up on empirical observation in biology – thinking in terms of design is simply a necessary device for interpreting what we see in nature, it does not involve a necessary commitment to a designer (although Kant is certainly willing to accede that it might). However, Ruse ultimately rankles at the belief that a particularly kind of thinking *has* to be done a particular way, while still acknowledging that if we could not ask 'what for?' then "biology as we know it today would be dreadfully impoverished" – and this is just as true in molecular biology, since "the genetic code is as much part of the design metaphor" as any biological feature.

Kant goes on to conclude that we can speak of "the wisdom, the economy, the forethought, the beneficence of Nature" but we don't need to make nature into an "intelligent being" since that would be "preposterous", and similarly we do not need to place an intelligent being "above it as its Architect" since that would be "presumptuous". Kant verdict on intelligent design, in other words, is that it postulates something more than is necessarily indicated. He is, in fact, extremely keen to keep talk of purpose out of science altogether, stating that we need to do this because the proper subject of science must be restricted to "that which we can so subject to observation or experiment" i.e. empirical research.

He also wrestles with the question of whether or not "all production of material things is possible according to merely mechanical laws." This explores whether it is necessary to invoke something more than mechanism (not necessarily God, but *something* more) in the explanation of how everything in the world came to be the way that it is. In the first part of the *Critique of Judgment*, Kant had made a distinction between different kinds of judgment he termed 'reflective' and 'determinant', and which correspond broadly to aesthetic taste in the first case, and empirical knowledge in the other. Using this distinction, Kant demonstrates that this issue comes out differently according to how we approach it.

From the point of view of aesthetics, he acknowledges that it is "a quite correct fundamental proposition" that "there must be thought a causality distinct from that of mechanism", namely "an (intelligent) cause of the world acting in accordance with purposes". However, this viewpoint applies *solely* from the perspective of our aesthetic experience of the world, and therefore from the viewpoint of the arts. From the perspective of empirical knowledge, Kant states that "this would be a hasty and unprovable proposition." Whatever our aesthetic sensibilities suggest, "we by no means undertake to concede reality" to something that is at root "a mere Idea". Whatever our instincts in this regard, the interpretation of the origins of the world is "always open to all mechanical grounds of explanation", and to follow where our sense of aesthetics leads would be to withdraw "beyond the realm of Sense into the transcendent" and possibly to be "led into error".

Kant explores all the major perspectives on this issue available in his time, and rejects them all. He grants to theism (the last interpretation he considers) one small concession, namely that "it certainly is superior to all other grounds of explanation" in so much as it "rescues in the best way the purposiveness of nature" from intellectual irrelevance by introducing "a causality acting with design for its production." Kant's point is that at least the theists offer *some* possible explanation for the appearance of purpose in nature, which no-one else at his time was really able to address. Yet this, he insists, isn't enough for the theistic interpretation to win out. In order for it to do so, it would be necessary to "prove satisfactorily... the impossibility of the unity of purpose in matter resulting from its mere mechanism". The idea of God as the origin of purpose in nature – the design argument – is not empirically valid in and of itself, and "can justify absolutely no objective assertion."

Once again, Kant says our aesthetic sense may accord with the idea of an intelligent designer, but we cannot validate this impression objectively. Nonetheless, he admits that we are "indispensably obliged to ascribe the concept of design to nature if we wish to investigate it." As noted above, we simply cannot think about organisms viably without recourse to the concept of design, even if we reject the idea of a designer as the explanation behind this. Adaptation is the modern secular descendent of the designer god in this sense: we have to talk about the features of animals in terms of design because that metaphor is irresistibly useful, but that doesn't mean we can talk about adaptation as a whole in terms of ultimate purpose – not without stepping outside of empirical science.

However, despite Kant's denial of intelligent design as empirically valid (i.e. as relevant for science), he falls considerably short of shooting down intelligent design *as a belief*. In fact, he suggests that the presence of natural purposes (i.e. organisms) represents "the only valid ground of proof for [the] dependence on and origin from a Being existing outside the world", a being who once inferred must also be intelligent, else such a being does not contribute to an explanation of the apparent presence of purpose in nature. He ultimately concludes that "Teleology then finds the consummation of its investigations only in Theology." Which is to say that Kant does not want to allow *any* role for the provision of ultimate purposes in science at all, and considers this a more suitable subject for theologians than for empirical research.

Kant reiterates this point by comparing the premature claim some might wish to make with all that we can reasonably assert in this regard:

If we expressed this proposition dogmatically as objectively valid, it would be: "There is a God." But for us men there is only permissible the limited formula: "We cannot otherwise think and make comprehensible the purposiveness which must lie at the bottom of our cognition of the internal possibility of many natural things, than by representing it and the world in general as a product of an intelligent cause, a God."

In effect, Kant says that we *must* think in terms of a designer (because we can think of adaptations in no other way), but this doesn't mean that there must be one *objectively*. This tallies nicely with contemporary perspectives on evolution, in which we continue to think of organisms as being designed, often subconsciously, but this kind of thought does not permit any de facto claim for the objective existence of a designer i.e. a necessary creator god. Conventional evolutionary thinking posits the designer in question as natural selection and related mechanisms – but interestingly, following Kant's reasoning, we should still question this deployment of an abstract idea – e.g. "Evolution" – in the role of designer. We can easily be misled by the lure of purpose-thinking into unwarranted conclusions.

This criticism applies to evolutionary beliefs as much as it applies to the Intelligent Design movement! The sum of our empirical knowledge of natural selection does not in and of itself validate claims of the form that such-and-such a feature evolved for such-and-such a purpose, even though the natural metaphor for talking about such things involves thinking of them *as if* they had been designed. Evolutionary perspectives deny the *intelligence* aspect of the origin of organisms, but they do not successfully evade the *design* part – they couch the appearance of designed features in terms of selection effects and so forth, but there is still in all such thinking an appeal to an ultimate purpose that is then used to evince explanations. How-Why games play the same way regardless of the justifying principle, and this should make us suspicious of all such claims when they are presented in ad hoc terms.

Kant's reasoning suggests something like a restraining order for keeping purpose-focused metaphysics out of science. If intelligent design must be excluded from science classes since it is not founded on empirical observations, then certain beliefs concerning the metaphysical implications of evolutionary theories are equally barred. If an ultimate explanation that culminates in God is not allowable for science, neither is an ultimate explanation that ends in something else – even natural selection. The jump to the conclusion that natural selection is the ultimate cause of everything in existence as is found, for instance, in Lee Smolin's (1997) concept of fecund universes (also known as cosmological natural selection) is not empirical grounded and to advance such a viewpoint is, Kant affirms, to take a *theological* position – albeit in this case one taking the 'anti' rather than the 'pro' God position. Neither viewpoint is strictly appropriate for a science curriculum.

Kant's restraining order suggests an equitable boundary on what science teachers may teach their students, irrespective of their own beliefs. Neither an intelligent being nor an abstract concept can be claimed to be the ultimate purpose or cause of life without stepping beyond empirical science. This means no intelligent design, and no adaptation or natural selection as ultimate causes of everything. This does not preclude talking about adaptation in terms of the metaphor of design, however, and there is no problem with invoking an intelligent manufacturer in archaeology, say, where talk of design is literal. All that is excluded are the more grandiose metaphysical claims, those which truly are untestable. As Ruse (2003a) suggests: "If it is science that is to be taught, then teach science and nothing more. Leave the other discussions for a more appropriate time."

As it happens, this solution has one rather giant snag: people don't agree *which* claims are untestable. For instance, Hugh Everett's (1957) 'many-worlds' interpretation of quantum mechanics, which postulates infinite parallel universes for which it seems there can be absolutely no evidence, appears to be an obvious case of metaphysics. However, Everett insisted his concept *was* falsifiable, and a surprisingly large number of physicists take it seriously – although it's far from clear what would falsify Everett's theory that wouldn't also falsify quantum mechanics as a whole, leaving it in a very suspect position. Similarly, the Intelligent Design movement considers its claims to be testable, and hence excluded from being considered metaphysics. What Kant has given us is a restraining order that we're going to find rather difficult to enforce.

Evolution Before Darwin

The *Critique of Judgment* was published in 1790; *On the Origin of Species* was published in 1859. Writing some seventy years before Darwin, you could be forgiven for thinking that Kant would have nothing to say on evolution, but this would be a mistake. The popular view that Darwin destroyed a prevailing belief in divine Creation is as historically confused as the popular view that Columbus destroyed a prevailing belief in a flat earth. (Both Columbus and his critics believed in a round planet, they merely disagreed as to its size – and as it happens, it was Columbus who was wildly in error in this case).

One quote in particular is frequently proffered with the intent of showing how wildly mistaken Kant was in respect of the prospects for understanding the origins of life (e.g. Zumbach 1984, Wolfram, 2002, Barah 2009):

It is indeed quite certain that we cannot adequately cognise, much less explain, organized beings and their internal possibility, according to mere mechanical principles of nature; and we can say boldly it is alike certain that it is absurd for men to make any such attempt or to hope that another *Newton* will arise in the future, who shall make comprehensible by us the production of a blade of grass according to natural laws which no design has ordered. We must absolutely deny this insight to men.

Sounds like a blunder, doesn't it? But Kant continues:

But then how do we know that in nature, if we could penetrate to the principle by which it specifies the universal laws known to us, there *cannot* lie hidden (in its mere mechanism) a sufficient ground of the possibility of organized beings without supposing any design in their production? would it not be judged by us presumptuous to say this?

In some respects, Kant was indeed mistaken – Darwin's natural selection *did* allow people to imagine organized beings arising from purely mechanical principles; this insight was not denied to humanity. But Darwin's theories *did not* rise to the explanatory power of Newton's laws; they did not mechanically explicate a blade of grass – they provided a new way of thinking about how an ordered design might emerge from natural laws. And this possibility Kant had afforded.

Indeed, Kant goes on at some length in the final sections of the *Critique of Judgment* concerning various issues that a naïve appraisal might be surprised to encounter seventy years before *On the Origin of Species*. Kant takes the idea of organisms changing – evolving – quite seriously, noting that many different kinds of animals show a common body plan such that "a great variety of species has been produced by the shortening of one member and the lengthening of another, the involution of this part and the evolution of that". He adds that this "allows a ray of hope, however faint" that there might indeed by a "mechanism of nature" that might explain the variety of animal life.

We can even see the seeds of Darwin's "descent with modification from a common ancestor" in Kant's text, since Kant notes that "this analogy of forms... strengthens our suspicions of an actual relationship between them", one that can be traced to a common parent and thus might allow a connection to be drawn "from man, down to the polype, and again from this down to mosses and lichens, and finally to the lowest stage of nature noticeable by us, viz. to crude matter." He thus notes that the whole of nature "seems to be derived from matter and its powers according to mechanical laws (like those by which it works in the formation of crystals)."

This isn't Kant pre-empting Darwin, but merely Kant expressing his views on the prevailing discussions in biology concerning the origins of life at his time, all of which was part of the background conditions of Darwin's work. In this regard, it can be quite surprising to uncover just how much had already been surmised. Kant comments that it is "permissible for the *archaeologist* of nature" to deduce "from the surviving traces" of the creatures that have existed on our planet that "mother earth" has given birth to various creatures which over successive generations show "greater adaptation to their place of birth". He considers this latter thought "a daring venture of reason" but acknowledges that "there may be few even of the most acute naturalists through whose head it has not sometimes passed."

Rather than Kant failing to anticipate Darwin, as is usually suggested, Kant was *already* thinking in terms of what Darwin was going to deliver – as indeed were a great many other thinkers at this time. What they lacked, and what Darwin delivered, was a viable mechanism i.e. natural selection. Darwin big idea was not, in fact, as explosively novel as is sometimes suggested, which is in no way to denigrate its contribution to natural science. Rather, Darwin's work appeared in the context of a scientific and philosophical world that recognized there were problems to be solved, and that was casting around for viable solutions.

However, despite these affordances, Kant – as Ruse (2003b)

puts it – considered that "the facts of nature go against evolution". His specific objections were in the first place that we have no experience of it, and secondly that it represents a problem for the concept of a self-organized being. If such a being produces others that are like itself, which (Kant affirms) is what we are used to experiencing, then we have to acknowledge a certain problem with deviating from that pattern. In some respect, this is almost Kant anticipating the neutral theory – recognizing that evolution not only had to explain the diversity of forms of life, but also solve the problem of how one can transform to another without destroying the self-organization so quintessential to living things.

The influence of Kant's thinking can be traced (as Ruse confirms) to the French naturalist Georges Cuvier, who first established extinction of species as a historical fact. Cuvier is practically the missing link between Kant and Darwin, his life chronologically fitting in between the two (he was born a few decades after Kant, and died a few decades after Darwin was born). Cuvier was strongly convinced by Kant's argument, and could not see how change could occur without disrupting an organism's ability to survive. Cuvier (1800) proposed a principle known as the *conditions for existence*, that suggests we can reason from the suitability of one feature of an animal to a particular lifestyle (e.g. the teeth of a carnivore) to the conclusion that the entire biology of the animal will be suited for that lifestyle (e.g. the digestive system of a carnivore will be suited to digesting meat). Cuvier wrote:

For it is evident, that a suitable harmony between those organs which act upon each other, is a necessary condition for the existence of the being to which they belong; and that if one of these functions were modified in a manner incompatible with the modifications of others, that being could not exist. (Reiss, 2005, Cuvier, 1800).

As Kant suggested, if it were not this way, the animal simply could not survive, and because of this, Cuvier (like Kant) found evolution exceedingly difficult to accept. Cuvier's original French phrase *conditions d'existence* is often translated as "conditions *of* existence", but Reiss (2005) provides good reasons for choosing "conditions *for* existence" instead: the former phrase is ambiguous, it may mean the necessary conditions *for* the existence of a particular animal, or it may mean the environmental circumstances *in which* it exists (which would include the nature of the other creatures living in that environment). The latter interpretation would be close to Darwin's fitness to environment, but this is not what Cuvier meant. He expressly intended the first interpretation, the one originally suggested by Kant, and thus Reiss proposes "conditions for existence" as the most accurate rendering of Cuvier's phrase.

Cuvier recognized that he was offering a purpose-focused or teleological principle, but it is one with a more restrained scope than many others offered for understanding the nature of life. For a start, it is essentially conditional in nature. As Reiss states: "If we observe that an organism exists, then it must be *possible* for it to exist, but this does not mean it was *designed* to exist, or that it had to exist." Reiss also suggests that Cuvier "appears to have believed in supernatural design", but was well-versed in Kant's critique of the design argument, which influenced him to accept that the only solid fact that explanations of the existence of a particular animal can rest upon is *the existence of that animal*. That much is undeniably objective, whereas claims to design (as Kant demonstrated) went beyond the facts.

This brings us to Darwin, who was influenced on the one hand by Paley's natural theology, with its perspective of looking at the living world as if it had been designed (as in the watchmaker analogy), and on the other by Cuvier, with the concept that the conditions for existence of any animal dictated a unity of function in its biology. With an insight born in part from fortuitous observation of the wildlife in the Galápagos islands, where subtly different species could be found living on neighboring islands, Darwin connected the design argument with Cuvier's functional understanding of biology and thus came up with his breakthrough metaphor of natural selection: the appearance of design could be linked to the conditions for existence, provided there is some permissible variation in animals between generations. Having seen it first hand in the Galápagos, he knew it had to be possible.

Although a silent partner in this watershed moment in the history of biology, Kant was still part of the dialogue that led to it – and this both in terms of discussion of the design argument, and in terms of influencing Cuvier's conditions for existence. Indeed, as Ruse suggests, Darwin's biology was practically Kantian in its nature – particularly in that it had no place for forces beyond physics and chemistry, yet still accepted an element of biological explanations that necessarily transcends physics and chemistry. The metaphor of design, uncovered by Kant, made its way into Darwin's theory, and from there, into contemporary science.

"The metaphor of design", Ruse states "is at the heart of the Darwinian evolutionary biology." Furthermore, understood in these terms he notes that "there is nothing very mysterious about purpose in evolution." It is because organisms have the appearance of design, because their conditions for existence require a particular way of being, that it is appropriate to talk about them in terms of function. Darwin's theory provides an explanation for why we find these apparently designed features in nature, but Cuvier's provides its context: if they were *not* suited for their conditions for existence, the creature could not survive and reproduce and thus would not be found in the world.

We need not be concerned about the metaphorical aspect of this approach, since as is abundantly clear at this point, there is no science without metaphors. Neither do we need to be concerned that thinking in terms of purpose when it comes to adaptations is somehow smuggling values into science (and not just because, no matter how strenuously scientists object, those values are always there somewhere!) Rather, it is appropriate to talk about adaptations in terms of imagined interests because it is by imagining those interests that we understand the feature. As Ruse suggests, "Kant was right in seeing that *we* do the science", *we* try to make sense of animals as if their biology had come about as a result of design and intent. It is us who imagine the feature as if it had been designed – if we didn't, we would have great difficulty understanding adaptations at all.

Returning one last time to the *Critique of Judgment*, the topic of intelligent design requires that we look at one final point Kant wished to make, which he did in a section he later relegated to an appendix. He had, through his various Critiques, systematically debunked each of the proofs of God that had been advanced as being inadequate (the disproof of the design argument for God we saw earlier, for instance). However, he advanced a different kind of 'proof' of his own – what Kant considered to be a *moral* proof of God. The short version of this argument is that it is rationally and morally necessary to attain the perfect good, and this is only possible if there is a God to secure an overarching moral order and causality. In the absence of God, we could *strive* towards perfect good, but it would be impossible to reach. This is not a 'proof' of God as a *fact* – it was precisely these kinds of proofs that Kant debunked. It is a *justification* for faith in God.

In exploring his moral philosophy, Kant's central idea was that humanity, via the power of reason, was capable of legislating moral laws for itself (an ability he believed, as a Christian, was God-given). In the *Critique of Judgment* he ventures that we may choose to believe that the existence of creatures capable of ethical reasoning could constitute "the final purpose of the being of a world" i.e. that humanity as ethical beings can be seen as the culmination of Creation (or evolution). He recognizes that we need not believe this, but notes that the alternatives are "either no purpose at all"(i.e. nihilism) or "purposes, but no final purpose" (i.e. subjectivism). Kant obviously favors the first option – the belief that humanity is the pinnacle of nature, that our ethics are our highest achievement, and that perfect good is attainable (because it is secured upon God).

Kant divided the kind of things we can mentally consider into three groups: opinion, fact and faith. Of matters of opinion, he had little to say. Matters of fact (or knowledge) he constrained to the faculty of reason and empirical observation of the world i.e. to science. But matters of faith were wholly separate from knowledge in Kant view. He stated that faith is "trust in the attainment of a design...the possibility of the fulfillment of which... is not to be *comprehended* by us." To put this another way, Kant says that even if we have faith in God, it is absolutely not on the cards that we should *understand* God's plan. He even admits "I cannot cognize what God is." Both God and God's plan are *incomprehensible* to humanity according to Kant.

If this is so, what value in the idea of God? Kant states:

If it be asked why it is incumbent upon us to have any Theology at all, it appears clear that it is not needed for the extension or correction of our cognition of nature or in general for any theory, but simply in a subjective point of view for Religion, *i.e.* the practical or moral use of our Reason.

Which is an incredible admission – for Kant not only allows that faith in God is *subjective*, but states that the only reason to contemplate God is for the ethical aspects of religious practice. Treating God as a matter of knowledge was a grave error as far as Kant was concerned – God should only ever be a matter of *faith*, and even then, the only way Kant allows that anyone can do God's work is to combine faith and reason in the pursuit of ethical living. This is a long way from the blind faith in Biblical authority that actuated the Intelligent Design movement, for he did not presume that "God's law" was static, factual and beyond question. Kant instead believed that reason has the power to *derive* an ethics of the universal that could potentially bring about a "realm of ends", the state of communal autonomy (Kant, 1785). Thus Kant believed that it was up to humanity to pursue what he claimed was our God-given purpose: learning how to live together.

Sneaky Theology

All this talk of theology may rankle anyone who feels that a book about evolutionary theories and biological science should steer clear of talk of God. However, this topic cannot be avoided while the contemporary debate concerning the metaphor of design involves the myth of intelligent design, and it will not suffice for those advancing other perspectives to simply demand that the Intelligent Design movement surrenders unconditionally. The kind of sneaky theology imputed to proponents of ID commenced long before that particular political activism was conceived, and there are good reasons for thinking that precisely what motivated ID in the first place was a kind of backdoor theological campaign with very different beliefs and goals.

It is worth remembering that while the roots of the conflict over intelligent design lie in previous centuries, the contemporary Intelligent Design movement *is not continuous with theological objections from Darwin's time*. As Midgley (2003) points out, the shock of Darwin's proposal wasn't that it was perceived as an attack on God – many of the proponents of Darwinian natural selection were Christians themselves. The real scandal concerned the dignity of *mankind*. Darwin treated the development of life as having been continuous, and this placed humanity into continuity with animal life, and thus "openly proposed to break down the fence that shut off our own species from other creatures." Darwin, as a consummate natural historian, recognized that this separation of man from animal was misleading and rather arbitrary – but for those who believed in the innate, God-given superiority of mankind (not to mention their own race and nation) this was too much to bear.

Thus while Christianity and Darwin's theory did come into conflict during the Victorian era, the argument was not being made that evolution disproved God. In fact, a great many Christians, who had (like Kant) been trying to understand the nature of life in the context of the ever-lengthening duration of the world's existence, welcomed the new understanding of life as an important missing piece of the puzzle of *how* God had made the living world. Darwin himself, while wavering on the doctrines of Christianity, retained close ties with his local parish church throughout his life, and by drawing on Paley's version of the design argument had instantiated his biological theory as essentially compatible with Christian theology. As Ruse (2003b) observes, Darwinism is effectively a child of Christianity and it is within families "that relationships are most intense, fraught and significant."

One of the chief reasons why the tale of insoluble incompatibility between evolution and Christianity has become so widely accepted is that the early Darwinians, and particularly 'Darwin's bulldog', T.H. Huxley, made this mythology central to proselytizing the new perspective on life. As Ruse puts the matter:

A major reason why the story persists of the ongoing fervent opposition to evolution – an opposition led and backed by Christians of all kinds – is that it suited the Darwinians to tell such a story. Huxley and his friends were striving honorably and successfully to reform Victorian Britain, and they needed a kind of secular religion to oppose the forces of reaction, as represented by the Anglican Church.

As a joke, Huxley once wrote to a friend about a biology course he was giving to school teachers "with the view of converting them into scientific missionaries to convert the Christian Heathen of these islands to the true faith" and as Ruse reports, the popular press referred to him as "Pope Huxley". There is a sense, therefore, in which evolution was offered as a secular religion *before* it had a chance to develop into empirical science. At Huxley's time, there simply wasn't an adequate methodology for exploring the claims of Darwin's hypothesis, which was little more than the metaphor of fitness to environment and a sketchy understanding of its implications. Thus in its early years natural selection largely *failed* as a science but, as Ruse puts the matter, "as a secular religion, it was a smashing success."

However, there is little direct connection between this vanguard of non-religious politics in Victorian Britain and the opposition to evolution that developed in the United States in the early twentieth century. In many respects, as Ruse contends, the rise of public education was the flashpoint of the conflict: conservative Christians in the heartland of the US were troubled by some of the ideas that were being taught in schools, ideas they "saw as not just false but as an ideology in its own right", one that originated with intellectual elites living in liberally-minded cities such as Boston and New York.

One of the things that struck me during my time living in Tennessee was that the intensely partisan politics of the United States conceals an essential conflict between its rural and urban citizens, and this operates on many levels. At the local level, politically liberal individuals living inside urban Knoxville (for instance) find themselves in frequent political conflict with conservative individuals living in the more rural areas surrounding the city. Both sides of this conflict have legitimate concerns about the relationship between city and countryside, but they are often unable to talk to one another clearly, in part because national issues tend to dominate political debate and local politics suffer consequent roadblocks and accusations of obstructionism on both sides.

At the national level, the same kind of division is echoed in the distinction between Democrat-favoring 'Blue States' – which tend to include major metropolitan areas that necessitate liberal views because of the diversity of ethnic groups living there – and Republican-favoring 'Red States' – the geography of which tends towards open countryside where people live with considerably more space between their fences than the typical New Yorker can ever dream of having. This distinction is arguably more important than religious differences, since the majority of US citizens still identify as Christians. It is the arrogance of Democrats to believe that their contemporary perspectives, forged in the melting pot of city life, must necessarily be taught to all citizens, and the arrogance of Republicans to believe the same of their more traditional beliefs. Dialogue seldom enters into the equation at all, and conflict is thus all too inevitable.

This backdrop serves as a partial explanation for why the school system has been the locus of legal battles in the United States over the last century, from the Scopes trial to the Dover school board. Presented solely in these terms, however, it seems as if those who argue against evolution are merely reactionaries, clinging to old beliefs in denial of the new facts. What this kind of story fails to recognize is that what is being opposed is not so much evolutionary science as it is a kind of sneaky theology that like Huxley's "secular religion" rides on the coat-tails of evolution, and dubiously cloaks itself in the authority of science.

Although by no means the only person guilty of this kind of excess, Richard Dawkins has been the most successful at upping the stakes of Huxley's game – he has not only offered an evolutionary non-religion, he has spread the message that evolution disproves God, and hence that *science* disproves God. Whatever one makes of his argument that the facts of evolutionary science remove any foundation for belief in God, the attempt to situate this discussion as a part of science is highly questionable. Any discussion of God in connection with science is *necessarily* theology, whether or not it concerns claims of God's existence or the contrary. Indeed, what Dawkins offers (alongside his often lucid accounts of evolutionary theories) is *atheology*, or even 'natural atheology', since it takes Paley's version of the design argument and turns it on its head.

In 1986, Dawkins published a book that was precisely a retort to Paley's watchmaker analogy, namely *The Blind Watchmaker*. Dawkins accepts the premise of the design argument – that the apparent complexity of the world is in need of explanation – and is even willing to accede to the inference that this complexity is evidence of something akin to design. However, he rejects the conclusion of the design argument, that the evidence of apparent design is proof of a *designer*. He suggests instead that if the apparent design of organisms is a parallel to a mechanical watch, then it must have been a *blind* watchmaker responsible for life i.e. natural selection, acting as a process but with no comprehension of outcome or purpose.

Rather than decompose the design argument into scientific and theological elements, Dawkins treats the whole thing as if it were a legitimate part of science. His argument rests on the idea that, on the lines presented in the watchmaker analogy, it is a legitimate scientific hypothesis whether the presence of design in nature can be attributed to a designer. Ironically, in this regard Dawkins is in agreement with proponents of intelligent design who *also* believe that the imputation of a designer behind the order of life is a legitimate scientific hypothesis. Even if this is accepted (and there are reasons we will explore in the next chapter for believing otherwise), the idea that disproving this hypothesis constitutes disproving God *isn't* a scientific conclusion – and this for the simple reason that *any* discussion of God is necessarily theology, and this cannot be taken as a legitimate part of empirical science.

It may be the case, as Dawkins asserts, that "Darwin made it possible to be an intellectually fulfilled atheist" – but atheism is not part of empirical science, but rather a part of theology (specifically atheology). This is a queer state of affairs to be sure, but it is inescapable that any stand on God – whether for or against – necessarily constitutes theology simply by definition of the word. Dawkins clearly feels very strongly that contemporary evolutionary theories *justify* his natural atheology, which is a matter of debate, and also that his conclusions are a part of empirical science, which is implausible. He is entitled to his beliefs on both accounts, but only for the same reason that any religious perspective is afforded latitude – namely that we have agreed to freedom of belief as a key value of contemporary civilization.

This *doesn't* mean that natural selection is barred from being a challenge to religious belief, of course, since religion was never construed to be a part of science in any ordinary sense. It just means than any discussion of these kinds of issues moves within the field of theology and not within empirical science. The claim, which Dawkins and others occasionally make, that questions concerning God can be legitimately explored by science is a wild transgression of Popper's milestone, a straying into the domain of metaphysics not for the purpose of establishing directions for future research, but for specifically atheological purposes. Not everyone is going to want to follow in this direction, whatever their feelings about God.

Dawkins mounts an even stronger atheological argument in his later book, *River Out of Eden* (1995). Considering the relationship between predator and prey via the example of the antelope and the cheetah, Dawkins asks the (theological) question: *if* there is a god that made both animals, what kind of god could it be? A cruel sadist? A dramaturge creating compelling natural history documentaries for eager TV audiences? He concludes that there is no possible account of a god compatible with this situation, the whole premise is ridiculous and thus there can be no ultimate purposes to life as a whole. According to Dawkins, attempts to construct religious meanings from life are necessarily false and misguided since if we look deeply into the universe we will find "at bottom, no design, no purpose, no evil and no good, nothing but blind, pitiless indifference."

As an atheist himself, Ruse is open to Dawkins argument up to a point, and suggests that faced with this kind of question "spokespersons in the science/religion community simply retreat". Ruse contends that any kind of satisfactory comeback to this kind of challenge must be on Dawkins terms – "adaptation, selection, blind variation, pain and all" and suggests that the only way for the religious community to respond is to admit that Dawkins is right – the Darwinian perspective *is* a serious challenge to religious belief. But, he adds that "once this is accepted, things start to fall into place" since there are theologies available that address these problems, and which offer a new perspective of God that take into account evolution. For Dawkins, this kind of move is intellectually dishonest; for Ruse, the matter is considerably more nuanced.

At the heart of the scientific beliefs that Dawkins holds (and which motivate his atheology) are two specific perspectives on the nature of life: firstly, that of gene supremacy discussed in the previous chapter, secondly, that adaptation provides the basis for the only viable understanding of the nature of life, a position usually termed *adaptationism*. An increasing number of researchers in evolutionary studies have accused advocates of adaptationism of misrepresenting the state of the field. Population geneticist Michael Lynch (2007), for instance, laments the way natural selection is sometimes invoked as an all-powerful force in biology without direct evidence. He suggests of adaptationism:

This stance is not very different from the intelligent-design philosophy of invoking unknown mechanisms to explain biodiversity. Although those who promote the concept of the adaptive evolution of [certain] features are by no means intelligent-design advocates, the burden of evidence for invoking an all-powerful guiding hand of natural selection should be no less stringent than one would demand of a creationist. If evolutionary science is to move forward, the standards of the field should be set no lower than in any other area of inquiry.

Lynch does not mention Dawkins by name, but in our private correspondence he has admitted that he feels Dawkins is "probably the biggest offender" and suggested that "blind adherence to natural selection as the one, and only, mechanism for evolution leaves no room for alternative views", despite the abundance of evidence that "this is an inaccurate view of evolution." The question is: has Lynch (and other scientists who feel similarly, for he is by no means the only researcher to feel this way) understood Dawkins position correctly?

Philosopher Peter Godfrey-Smith (2001) offers an intriguing attempt to unravel the dispute about adaptationism, and begins by recognizing that the way the term is used conflates *three* different meanings. Firstly, there is the kind of adaptationism savaged by Lynch that offers adaptation by natural selection as the only mechanism needed to understand evolution. Secondly, there is the kind of adaptationism that doesn't deny other mechanisms acting in evolution, but which believes that the important or relevant aspects of biology are explained by adaptation – that adaptation solves the "most important problems". Finally, there is the more modest claim that looking for features that can be explained by adaptation and the metaphor for design is simply a good method for evolutionary studies.

Interestingly, Godfrey-Smith denies that accusations of the

kind Lynch advances apply to Dawkins. He suggests that Dawkins *doesn't* believe natural selection is the ultimate and only cause needed for explaining biology at all, and rather that the kind of adaptationism Dawkins and philosopher Daniel Dennett espouse is the second kind – that which claims adaptation is the answer to the "big questions" in the evolution of life. However, he also admits that Dawkins and Dennett occasionally write from a "more ambitious position", sometimes expressed by criticizing those who seek explanations in terms of something other than selection, or by making a specific claim as to the extent the biological world has been shaped by selection. In so much as they occasionally slip up and seem to adopt the grander claim, Godfrey-Smith allows that criticisms of Dawkins and Dennett in the style of Lynch's are entirely justified.

However, there is a deeper problem in Dawkins' kind of adaptationism that Godfrey-Smith is keen to bring to light, namely that while many might agree that the appearance of design in nature is interesting, "this is apparently just a fact about us." He denies, therefore, that there are objectively "most important" biological questions at all, seeing this instead as a subjective individual choice. This being so, Dawkins and Dennett's adaptationism "is revealed to be little more than the personal preference of some biologists and philosophers; they find selection important because it answers questions that they happen to care about." It is not a *scientific* view at all, but a belief *about science*.

Godfrey-Smith's use of "personal preference" may strike some people as too weak a description for the way Dawkins and his friends discuss evolution. Lynch, for instance, suggested to me that this term "implies that they have given serious consideration or even realize that there are alternatives". In his view, as noted previously, the kind of deep commitment to adaptationism which he sees Dawkins exemplifying is a commitment equivalent in some respects to the purportedly religious-motivations behind the Intelligent Design movement. To some extent, Godfrey-Smith does recognize that Dawkins goes beyond science in his arguments, suggesting that "the efficacy of evolutionary replies to theological arguments is extrinsic to the scientific work done by biologists themselves". He states:

Dawkins sees apparent design as the one thing that, before Darwin, could rationally motivate a traditionally religious outlook on the natural world. Darwin, by destroying the Argument from Design, thereby reshaped the entire intellectual landscape. The concept of natural selection is a pin holding much more than evolutionary biology in place; it is holding together the scientific/enlightenment world view.

However, as far as Godfrey-Smith is concerned, the task of relating biological theories to religion "belongs primarily to philosophy of science." Thus, the philosopher defends Dawkins in so much as he suggests that many of the criticisms leveled against him are based on the assumption of a strong form of adaptationism that Dawkins doesn't really believe (but occasionally lapses into). He additionally proposes that Dawkins is frequently misunderstood in the context of genes, suggesting that once again what Dawkins asserts in connection with genetics is "a claim about the significance of genetic influences on behavior". Godfrey-Smith sees this less of a claim to genetic determinism and more of a "philosophical error".

Other critics are less generous when considering Dawkins' beliefs. The scientist often regarded as the founder of the field of theoretical biology, Brian Goodwin, called Dawkins "something of a preacher" (Brockman, 1995). He sees the belief system that Dawkins holds to as a kind of transformed Christian fundamentalism. Dawkins claims that organisms are constructed by genes, which gives rise to the metaphor of the genes as selfish, which in turn is reflected in the 'selfish' interactions between competing organisms, but we can escape from this inheritance and become truly altruistic. For Goodwin, this is equivalent to the classical Christian doctrine that humanity is born in sin, we have a selfish inheritance, we are thus condemned to conflict, but that salvation is possible. He states:

What Richard has done is to make absolutely clear that Darwinism is a kind of transformation of Christian theology. It is a heresy, because Darwin puts the vital force for evolution into matter, but everything else remains much as it was. I suspect that Richard was at one stage fairly religious, and that he then underwent a kind of conversion to Darwinism, and he feels fervently that people ought to embrace this as a way of life (Brockman, 1995).

It seems that something close to Goodwin's account is a fair way of understanding the circumstances behind Dawkins' beliefs. Indeed, Dawkins himself has noted that he was raised as an Anglican but had doubts in his teenage years, stating that the main reason he retained his Christian belief was that he was "impressed with the complexity of life" and felt that it "had to have a designer". When he recognized that Darwin's account was a "far superior explanation" it "pulled the rug out from under the argument of design". Christianity became impossible for him at that point (Hattenstone, 2003).

But perhaps by starting out as a staunch believer in the design argument, Dawkins has ended up unable to fully rid himself of his theological roots – he may have switched to atheology, but he is still one of the louder voices in modern theological discussions. Haught (2010) suggests that Dawkins and Dennett "operate as cryptotheologians by insisting that natural selection is a substitute for the traditional theological accounts". They may explicitly reject God, but "the ghost of theology lives on in their hunger to find an *ultimate* (metaphysical) explanation of design in evolutionary biology alone." He is in the strange situation of "playing his game in a theological arena" while simultaneously insisting that "theologians must be disqualified from appearing there."

As Haught (1995) attests, "the days in which scientific ideas could be used to seal arguments for God's existence are over." Yet despite the collapse of the traditional theological design argument – as foreshadowed hundreds of years earlier by Kant – the teaching of science as a subject is still infiltrated by sneaky theology from both proponents of intelligent design *and* enthusiastic atheological advocates of adaptationism on the other. Kant's restraining order is not being enforced by anyone. As Haught (2010) states, "science and science education have become the victims of metaphysical exploitation by both sides." Until discussions of biology can occur without reference to ultimate questions, biology and theology will be unable to successfully part company.

Beyond Adaptation

The metaphor of design provides a new way of understanding adaptation in nature, and helps to demonstrate how much of the intelligent design debate involves issues lying beyond Popper's milestone – as well as showing that this applies to *both* sides of this argument. However, the problems associated with adaptation go beyond the myth of intelligent design – there is also the myth of adaptationism that Godfrey-Smith critiqued, which applies explanations in terms of the metaphor of design further than is warranted – something even Darwin fell prey of, following the design argument into "too great a conviction that adaptation rules supreme" (Ruse, 2003b). This issue, which entails questioning whether How-Why games can ever be legitimate science, requires some further consideration. Is the way forward with adaptation to clarify when it is acceptable to talk in terms of the metaphor of design, or to move beyond this perspective entirely?

It is worth clarifying what the term 'adaptation' is meant to imply. George C. Williams bears much of the responsibility for putting emphasis on the term, and central to *Adaptation and Natural Selection* (1966) was the idea that genuine adaptations had to be distinguished from "fortuitous benefits" or (which is to say the same thing) "incidental effects". As Sober and Wilson (2011) observe in this regard:

> Fortuitous benefits are not *adaptations*, though they are *adaptive*. Adaptation is a concept that looks to the past; to say that a trait is now an adaptation is to make a claim about its history. Being adaptive (or advantageous) is a concept that looks to the future; to say that a trait is now adaptive is to say that it promotes survival and/or reproductive success. The title of Williams' book was well chosen – adaptation is a concept that is fundamentally linked to the concept of natural selection.

Much of the debate about adaptationism revolves around distinguishing between what is or is not an adaptation in these terms, and the denial of any claim that treats *any and all* biological features as having come about by a process of adaptation. The most famous attack of this kind came from Stephen Jay Gould and Richard Lewontin (1979). Considering the architecture of San Marco, Gould and Lewontin suggest that certain features occur not as a necessary part of the design of certain buildings (as in the case of load-bearing columns, without which the construction would collapse) but as secondary consequences of the architectural configuration that results. In particular, highly ornate spandrels appear at the intersecting cornices of many Renaissance cathedrals, but these were added to take advantage of a physical space that in itself occurs solely as a consequence of the necessary constraints involved in putting together buildings of this kind.

By analogy, they suggest that certain features in biology cannot be *known* to be adaptations – the equivalent of a load bearing beam – and might instead be the equivalent of the architectural spandrels. Indeed, they coin the term *spandrel* in direct reference to this, as a term meaning any feature of an organism that has appeared without specific adaptation having occurred. They further advocate caution in assessing just what qualifies as an adaptation, as opposed to a spandrel. Without access to the evolutionary history of a creature (which we never have), can we really be sure what constitutes an adaptation?

This paper, as Godfrey-Smith notes, challenges not just the ambitious form of adaptationism that seeks to explain all biology in terms of adaptation but even the more modest proposal that views explanations in terms of adaptation as a useful method for conducting biological research. Gould and Lewontin argue that there is a tendency to go far beyond scientific thinking when adaptation is being discussed – and further suggest that thinking in these terms encourages attempts to rescue a hypothesis that appears to have been disproved by simply invoking an ad hoc explanation. In effect, this is a criticism of anyone who resorts to How-Why games in the absence of any empirical data that might be used to ground them.

However, it should not be thought that Gould and Lewontin are seeking to deny adaptation, as such. As Lewontin (1978) has stated, adaptation "is a real phenomenon". The fact that "fish have fins, that seals and whales have flippers and flukes, that penguins have paddles and that even sea snakes have become latterly flattened" is not a coincidence but evidence that the problems of moving in water have "been solved by many totally unrelated evolutionary lines in much the same way." Lewontin allows that "it must be feasible to make adaptive arguments about swimming appendages", even if he (with Gould) expresses skepticism about how far these arguments can be pushed.

The funny thing about the incredibly heated debate that grew up around this idea of spandrels is that deep down, no-one inside the scientific community that discusses evolution seems to feel that the kind of strong form of adaptationism identified by Godfrey-Smith has any followers. As Herre and colleagues have observed (2001), if what's meant by this term is the idea that all animals are "perfectly adapted in all respects for all situations that they encounter" then this thesis is "a ludicrous proposition and not worth testing". The evidence of biology runs wildly counter to this claim, and as such, no biologist worth their salt could advance it. Strange, then, that the spandrels paper should have been so controversial: if this isn't the crux of the debate, what is?

Ruse suggests that the problem goes to the way that adaptation is used, suggesting it is the paradigm (explicitly refer-encing Kuhn's term) by which biological science examines organisms, and that it is impossible to simply set this aside. Raising concerns about applying this methodology too widely is one thing, but denying the paradigm entirely is another. He suggests that "it is because things do work so well so much of the time that we can feel justified in the adaptationist paradigm and can so profitably seek out and find explanations for the excep-tions." He insists this is a strategy, "not a blind metaphysical commitment to adaptation without exception." Furthermore, he challenges Lewontin and his followers to offer a serious alter-native: "When they demonstrate that they can do science which explains and predicts without invoking adaptation even implicitly, then we can start to take their position seriously."

In the way Ruse mounts his defense against Lewontin, he explicitly characterizes the spandrels argument as a claim that "we should not play the game at all". I'm sensitive to these objec-tions – after all, as Kuhn himself made clear, it's not possible to simply abandon the current paradigm; some alternative must be presented, and Gould and Lewontin's advice to be cautious doesn't quite rise to this level. However, I see Lewontin's position in this regard as essentially deflationary: it concerns what can or cannot reasonably be deduced within science. What is all too often being played are How-Why games that are indistinguishable from those played by Victorian scientists in a theistic manner. Ruse and others may be careful to avoid this, but these kinds of mistakes constantly occur.

Molecular biologist Michael Lynch (2007) presents a more detailed critique of the problems relating to adaptationism than the spandrels objection. Lynch is an absurdly prolific scientist working on a variety of significant problems in population genetics, and like Lewontin, he recognizes that adaptation does occur in many instances. Also like Lewontin, he thinks that adaptationism all too frequently goes beyond what is justified by contemporary science, especially in so much as mechanisms other than natural selection are all too often ignored. Crucially, whereas others speculate about abstract features such as complexity, modularity, and capacity for evolutionary change, Lynch suggests (in a spandrel-like manner) that these "may be nothing more than indirect by-products" of various processes that occur at the level of the cell and the gene. The origins of the diversity of life, in Lynch's view, "have roots in nonadaptive processes".

Lynch identifies a long list of claims he calls myths, meaning in this case assertions that no longer have sufficient support to be justified i.e. factual errors. First and foremost, the idea that evolution can be understood solely and primarily as natural selection. Rather, Lynch suggests there are at least three other key forces that affect evolution, none of which are related to claims to fitness of any kind. Firstly, the mutations that provide the source of variation upon which natural selection acts, without which selection of genes could not occur. Secondly, recombination of genetic material, which shuffles the genetic deck in ways highly significant for the kind of population genetics Lynch studies. Finally, the kind of genetic drift identified by Kimura which results in a variation in the frequencies of gene variants from generation to generation, irrespective of other influences.

Lynch calls these "the fundamental forces of evolution", although this strikes me as overreaching. Yes, natural selection is not the full story of evolution, but neither is population genetics a complete account of the history and nature of life. As explored in the previous chapter, symbiosis has played an incalculably important role in the story of how life on our planet came to be the way it is – and this cannot simply be collapsed into population genetic explanations. As a molecular biologist, perhaps Lynch's restricted viewpoint in this regard is understandable (although he does offer a brief nod to Margulis' theories, if not to her), but we should certainly be careful not to let the critique of adaptationism lead us back into gene supremacy.

Fortunately, Lynch's understanding of his own field shows a much richer comprehension of the issues concerning the connection between genetics and evolution. One of his myths is the idea that finding genetic differences between animals is sufficient for identifying the mechanism of evolution – on the contrary, he suggests that while changes at the molecular level are critical, all we can do is identify the "end products of evolution", not the "population genetic processes that promoted such change". He remains adamant that population genetics is still central to evolution however, and one of the myths he denies is the idea that the evolution of complex animal forms cannot be explained by changes in the frequency of gene variants.

Many of Lynch's myths concern claims made about natural selection that he contends are false. Most crucially, he denies that "natural selection is the only force capable of promoting directional evolution", pointing out that mutation and other genetic mechanisms have non-random effects that significantly drive patterns of evolution when the size of the breeding population is sufficiently small. Similarly, he dismisses as a myth the claim that mutation's only role is to create the variation that drives natural selection – on the contrary, Lynch counter-intuitively claims that mutation is in itself "a weak selective force" that can eliminate certain gene variants, providing strong experimental support for his claims.

Neither complexity, the genetic modularity of regulatory structures nor the ability to evolve, claims Lynch, can be justifiably afforded to natural selection. In the case of complexity, the opposite seems to be the case since there is "substantial evidence" that when selection is less efficient, complexity can increase. In the case of 'evolvability', Lynch states that there are no abrupt transitions between the genetics of single celled organisms and their multi-cellular brethren, as is sometimes claimed. Whatever adaptationism might suggest in this regard, if there is a selective advantage to complexity of organisms "one can only marvel at the inability of natural selection to promote it" since, as Margulis often notes, the single celled creatures still dominate the planet in terms of numbers, biomass, and species diversity.

What's more, in perhaps the most damning indictment of the focus on natural selection in the evolution of life, genetic drift is not merely a random process producing 'noise' but otherwise leaving adaptation to reign supreme – Kimura's discovery serves to reduce the efficiency of selection and thus can increase the chance that beneficial genes will be lost and that deleterious mutations will persist. Far from being a passive irrelevancy to evolutionary theory, genetic drift is highly significant. Indeed, as the myth of the flexible gene has already revealed, adaptation and nonadaptive genetic evolution are inseparably linked.

In an echo of the spandrel concept, Lynch states that "many aspects of biology that superficially appear to have adaptive roots almost certainly owe their existence in part to nonadaptive processes." The justification for playing adaptive How-Why games with the metaphor of design is wildly undercut by what has been learned by molecular biologists in recent decades such that "there is no longer any justification for blindly launching suppositions about adaptive scenarios without an evaluation of the likelihood of nonadaptive alternatives." Nonadaptive processes "have played a central role in driving evolutionary patterns", to the extent that it may no longer be reasonable to think in terms of Dawkins' metaphor of "climbing Mount Improbable". Rather, much of what has occurred may be expected consequences of the features of DNA as a mechanism for inheritance. In so much as this claim can be verified, adaptation ceases to be the central player in the story of life, and must instead be understood as part of an ensemble cast.

Biologist John O. Reiss (2005, 2009) takes these kinds of arguments further than any of his contemporaries by mounting an attack not just on the myth of adaptationism, but on the paradigm of adaptation itself. In an extraordinary about face, Reiss takes the argument back to the conditions for life espoused by Georges Cuvier, and hence to a standpoint remarkably close to what Kant offered in his *Critique of Judgment*. Just as Lynch denies the claim to improbability offered in evolutionary histories, so Reiss savages the idea that organisms can be understood as sets of independently optimized traits. This approach, according to Reiss, is reliant on the analogy between creatures and artifacts that was inherent in the design argument, and still persists in the metaphor of design. As David Depew (2010) comments in respect of Reiss work, adaptationists of all stripes rely on models of optimization that "presuppose that natural selection is a counter-entropic 'force' that drives populations up the side of what Dawkins revealingly calls 'Mount Improbable'". Reiss sees this approach as thoroughly misleading.

Reiss recognizes the metaphor of design that Ruse advances, but sees it "as a serious problem." As he remarked to me, the metaphor of design causes trouble because it "obscures or confuses features of the phenomenon one is studying" leading to "some pretty strange views of the evolutionary process" that are "particularly dangerous here because we don't have the rigor to go along with the metaphor" (Reiss and Bateman, 2011). The nub of his complaint is that treating natural selection as the designer in the contemporary update of the design argument is too literal an interpretation of the evolutionary process. Instead, he suggests recognizing the conditional aspects of the story behind each creature's circumstances – in particular, the fact that any particular animal is present in the world is itself the strongest available evidence for its evolutionary history: given that the animal does indeed exist, we can explain that existence in terms of the conditions necessary for that circumstance to have come about (Reiss, 2005).

Cuvier's conditions for existence are precisely this kind of conditional explanation – they are not, Reiss admits "terribly informative", but this is compensated for by the dual benefits of being complete explanations within the context under consideration (why an animal exists) and being decomposable into smaller elements, allowing them to serve as useful functional explanations (how an animal has acquired certain capacities). The argument for taking a Cuvier-style approach to evolution isn't that it is superior in explanatory power to the metaphor of design, but rather that perfect explanation of organisms is entirely "beyond our ability", as Gould and Lewontin's spandrels critique made clear. We can't know the evolutionary history of any animal with significant degrees of confidence, but we can consider explanations in terms of function (as Cuvier did). If we constrain our interest to observing some result and then explaining it in terms of the conditions that made that possible, we are radically less likely to be driven into strange How-Why games or other excursions out beyond Popper's milestone.

What Reiss proposes is using Cuvier's *conditions for existence* as the boundary conditions within which evolution must have occurred – effectively supplanting the mythology of adaptationism entirely. Since creatures that were not adapted to their circumstances necessarily died, there is a sense in which the conditions for existence is identical with the principle of natural selection, albeit a rather broadly construed version that is severed from its usual connection with the metaphor of design. When we approach life in this way, the perspective on adaptation is very different from what we are used to since it "is not a product of evolution, it is a condition for evolution." Every animal alive today must be able to trace a chain of inheritance through ancestors all of whom were adapted to the conditions for existence that applied during their lifetimes. Adaptations, in Reiss' point of view, are not some incredible deviation from probability, but the accumulated life history of billions of conditions for existence, the advantages of which have been passed on to the various descendents.

However, Darwin's concept is not without a role in Reiss' approach. In order for any trait to continue to exist, the creatures who express that trait (and also the genes that allow that trait to occur) "must survive and reproduce at a higher rate than alternatives" precisely because that trait provides advantages to that particular animal. (This is a narrow sense of natural selection that compliments the broader context of the conditions for existence mentioned above). This is the contribution that Darwin and Wallace offer that extends beyond Cuvier's principle, but it does not (according to Reiss) "give us a reason to compare natural selection to a designer." Instead, it invites explanations of traits in terms of the conditional circumstances that lead to their occurrence – and he insists this is "what most successful evolutionary explanations have been."

The whole purpose of discussing the evolution of life in terms of conditions for existence instead of the metaphor of design is a pragmatic attempt to avoid the kind of confusions about the design argument that have been detailed in this chapter. "Nature doesn't aim at anything," Reiss asserts, "and thus can't miss." However, from the point of view of any given animal, there is "an explicit welfare component" that is essential to any attempt to explain a biological function, since "overall adaptedness is something that must, by definition, have been maintained." This is to say, in terms of my myth that advantages persist, the very fact that an animal exists is the evidence that its biological traits were beneficial to its survival. If they had not been, it would be inconceivable that it or its ancestors would have endured, and it is this survival – its continued existence – that we are attempting to explain.

Reiss admits that in his attempt to clarify the confusions concerning the metaphor of design, he may accidentally have "introduced a new source of confusion into the ongoing debate." Personally, I would suggest this debate is already so inconceivably convoluted that I'm not convinced it is possible to add any further confusion. However, while I find Reiss' approach to be a fascinating alternative to adaptationism – and not least of all because it returns to a Kant-inspired conception of life that predates contemporary theories of evolution while still recognizing what Darwin's metaphor added to our understanding – I am not yet convinced that the metaphor of design has run its course.

I challenged Reiss to consider whether or not abandoning the metaphor of design would make evolution more complex and obscure, and thus difficult to teach (Reiss and Bateman, 2011). He remains optimistic that the kind of view from beyond adaptation he offers might actually be *easier* to grasp than those evolutionary stories descended from the design argument, noting a "certain unreality" in the way natural selection is presented, making it "harder to understand":

Evolution needs to be presented based on evidence for its reality... Once the fact of evolution has been established, then the basic mechanism of mutation and differential survival and reproduction can be brought in to explain it. All of the talk of natural selection and adaptation just obscures. However, the chance that I will prevail in this argument is rather small – too many are wedded to the current paradigm.

Between Lynch's rigorously presented criticisms of the genetic basis for adaptationism, and Reiss' pragmatic account of the conditions for existence that finds its roots in Cuvier and Kant, we face an interesting situation. The Intelligent Design debate is, to a certain crude degree, between the myth of intelligent design on the one hand and the myth of adaptationism on the other. Yet neither of these stories seems likely to be a part of our understanding of evolution going forward. Whether we accept the metaphor of design espoused by Ruse, the conditions for existence offered by Reiss, or (as I am drawn to) a middle path that accepts the merits of both these new myths in terms of reimagining the history of our planet, our perspective on the nature of life will never be quite the same again.

The swift's wings can be compared to those of a fighter jet by the metaphor of design: just as an engineer designs an aircraft with aerodynamic efficiency, so the functional features of a bird's wing show an extraordinary fit to its lifestyle as an agile hunter of aerial insects. This is an incredible aspect of life, but it is not an improbable or unexpected occurrence. Rather, it is the almost inevitable consequence of the chain of inheritance passing advantages through generation after generation of animals, each of which was necessarily fitted to its own conditions for existence. Those species that failed to remain within these metaphorical boundaries are no longer with us, and that is one of the things we call natural selection. That these conditions for life have gradually allowed for ever more subtle and sophisticated capabilities is what we call evolution. It is the nature of life that the delicate counterpoint between environment and animal, between survival and death, between continuing the chain and otherwise, gives rise to the incredible diversity of creatures on our planet, and passes to each those gifts its ancestors needed to prosper.

5. True Myths

Fact and Fiction

The story of the myths of evolution I have presented in this book has hopefully revealed the remarkable importance of metaphor in science, particularly in biology and the study of the history of life. I have said throughout that these myths can be thought of as metaphors or as imaginative fictions, but so far I have not pressed this idea that science, which we conventionally think of in terms of fact and truth, is intimately dependent upon fictions. How is it that we get truth from fiction?

We are accustomed to thinking about 'fact' and 'fiction' as entirely contrary – something can't be *both* fact and fiction, right? However, it is not so easy to draw a line between the two. Historical fiction, for instance, blurs the distinction, and as we've seen throughout this book there are many cases where a metaphor – which is a kind of fiction – can be extremely useful in understanding matters of fact. Rather than thinking of fact and fiction as opposites, it might be helpful to recognize that there is a sense in which fact and fiction are variations on a theme. I don't mean to suggest there is no such thing as truth, but I intend to demonstrate that even the truth is a *kind* of fiction.

News reports provide a vivid example of this relation between fact and fiction. Suppose you watch footage of a hurricane battering the southern coast of the United States on a news program. We would certainly want to call this film factual. Yet the same footage could appear as an establishing shot in a movie, perhaps in a story where a fictional hurricane plays a role in the plot, and we would certainly want to call *this* film fiction. But it's the *same* footage in both cases – the only thing that has changed is the *context* in which it is presented. It seems that it is not that something is fact or fiction in and of itself, but rather that there are ways of presenting things *as fact* (such as a news show) and ways of presenting things *as fiction* (such as a movie).

The philosopher of art, Kendall Walton (1990), has developed an influential theory for understanding fiction he calls the make-believe theory of representation. According to Walton, fictional truths can be implied by various objects (which he terms 'props') by virtue of their properties and the cultural circumstances within which they are interpreted. So, for instance, it is a fictional truth about the Statue of Liberty that it is a woman, or, as Walton prefers to state this, it is *fictional* that the Statue of Liberty is a woman. Remember, the Statue of Liberty is *really* copper sheeting shaped to resemble a female figure – it is not *really* a woman, it only *represents* one. So it is true that the Statue of Liberty is a statue, and that it is made of metal, but it is fictional (that is, fictionally true) that it is a woman.

What's more, we can talk about fictional matters *as if* they were true, by assuming the pretence that what we are talking about is real. We do this all the time, in fact. If we say "when the Statue of Liberty was dedicated, she was the tallest thing in New York" we are accepting the pretence that it isn't just fictional that the Statue of Liberty is a woman, hence we can talk about the inanimate statue as 'she'. It is incredibly natural for us to talk this way. We normally wouldn't think twice about saying "Robinson Crusoe was marooned on an island" when what we really mean is "in the fictional book *Robinson Crusoe*, the title character is marooned on an island". Talking about fiction as if it was fact is so comfortable we often don't even notice the pretence: when a fan of the New York Mets calls them "the best team in the world", this definitely isn't based on the facts of their performance – it means something like "in the pretence I accept as a fan of the Mets, it is fictional that they are the best team in the world".

Walton, drawing on the work of Stanley Fish, suggests that what counts as fact or reality depends upon which stories have been *authorized* as true. This doesn't mean that nothing is 'really' true, and it certainly doesn't mean that truth isn't important. Walton notes that "truth and reality, whatever they are, obviously *do* matter." The point about *authorized stories* is to draw attention to how we reach our conclusions about the difference between fact and fiction. News reports, legal findings, scientific conclusions and so forth all have a certain authority associated with them – drawing, in these three cases, from the trust we place in journalistic integrity, the rule and process of law, and in the scientific method respectively. If we overhear a rumor, it lacks authority – when the same story appears in a respectable newspaper, we become comfortable accepting it as true.

This idea of an authorized story is relevant to the case of the hurricane footage. When this appears in the movie, we *imagine* that there is a hurricane since it is fictional in the world of the film that there is a hurricane. If the same footage is presented *as news* it gains the authority to be believed as true *as well as* imagined. The images we see and the sounds we hear are the same in both cases – the only difference is whether there is any source of authority that supports its claim to being considered fact. Walton says "what is true is to be believed, what is fictional is to be imagined" but I would say "whether it is fact or fiction, it is to be imagined; if it is authorized as fact, it is *also* to be believed true".

What I find fascinating about news programs is just what is authorized as fact: there is 'news' i.e. stories about death, injury, violence, disaster, crime, war, crowds, anger, money, power, technology and knowledge, but also 'sport' and 'weather'. We do not think twice about this, yet sport is reporting the events of sporting matches as both factual and significant, that is, lending authority to what would otherwise be 'just a game'. Similarly, the weather forecast is authorized as fact, even though it is obviously far closer to fiction (as anyone who has tried to rely upon the Met office for an accurate prediction of the British weather will attest!).

Thus the only distinction between fact and fiction is that what we accept as fact is authorized to be believed as true, *as well as being fictional*. 'Fictional' in this sense means that we imagine the content of the story (whether or not we consider it to be true or real), and 'factual' means that in addition to imagining it, we imagine that it is real. This is an unfamiliar way of looking at truth, but it has the benefit of showing how the act of imagining is something that *both* fact and fiction share in common. Rather than fact and fiction being polar opposites, fact is simply a *different* kind of fictional story, one that has the authority to make a claim to truth. Fact is not the opposite of fiction, but a different kind of fiction.

When it comes to science, we frequently run into problems because we – quite understandably – afford considerable authority to the research that scientists conduct, in part because empirical testing is generally more robust than mere opinion. We place trust in the claims of scientists, but we don't often look at the fiction involved in science. As I have tried to show throughout this book, science is chock full of fiction in the form of myths and metaphors. Often these help us to understand the facts under consideration, but we have to be careful not to become confused about how far that scientific authority extends. The belief that 'scientific' is a synonym for 'true' cannot be sustained any more, and not just because of Kuhn's critique. So much of what is true comes to us only by its relationship with fiction. The old image of truth and fiction as opposites cannot be sustained – we need a new mythology that allows for *truth from fiction*.

In my previous philosophy book, *Imaginary Games* (2011), I detailed the foundations of this approach, drawing on Walton's make-believe theory of representation and also in the final chapter on Stephen Yablo's adaptation of this approach to mathematics, showing that even numbers are essentially metaphorical. I will not reiterate this content here, but I will summarize my conclusion: fiction is intimately involved in all aspects of our lives, from the arts to ethics to science. As Reiss acceded to me when I first challenged him on his choice of the conditions for existence metaphor in preference to the metaphor of design:

Metaphor certainly helps the public understanding, and often plays a role in scientists' conceptual understanding – for example one might compare gravitational to emotional attraction. In a way mathematical models are themselves a sort of metaphor for what is "really" going on (Reiss and Bateman, 2011).

Of course, science is not the only field with such an intimate relationship with fiction and metaphor: religions of all kinds make use of both kinds of imaginative thought, and this is true whatever we believe. Even the most orthodox Christian cannot deny that Jesus' parables are fictional stories told to make spiritual and ethical points, nor that when John the Baptist declares Jesus to be "the lamb of God" that this is certainly a metaphor (no-one believes Jesus was literally a bleating, woolly sheep). Although religion and science are obviously different kinds of activity, they share in common a central role for imagination and metaphor, as Catholic theologian John F. Haught (1995) observes:

Both science and theology generate imaginative metaphors and theories to interpret certain kinds of "data," but in neither case is it always clear just where metaphor or theory leaves off and "fact" begins. Indeed the consensus of philosophers today is that there are no uninterpreted facts.

Religions, of course, are the traditional mythologies of the world – and by this, I do not mean that all religions are false. As the folklorist Joseph Campbell (2001) observed, myths serve a vital role in conveying the ethical truths of religions. He was keen to stress, as I have, that myth should not be taken to mean 'false' – the belief that religious fiction (like Jesus' parables) are necessarily false results from a misreading of the nature and purpose of metaphorical language. Indeed, Campbell (1972) not only states that all mythology is inherently metaphorical, but that this symbolic aspect often goes unnoticed by religious practitioners since it has "always been the way of multitudes to interpret their own symbols literally", that is, to confuse fiction and fact, metaphor and meaning.

What Campbell calls a *living mythology* is a set of stories and myths that have an active role in a culture, serving "to validate, support, and imprint the norms of a given specific moral order, that, namely, of the society in which the individual is to live." Katherine Hayles (1995) makes similar claims about what she terms *grand narratives*, which are "the stories a culture tells to make sense of the world and itself." Such living mythologies or grand narratives have always been central to the way humans have lived, and they continue to serve this role since, as Mary Midgley (1991) has suggested: "To make myths is to express through symbols one's most profound, fundamental beliefs – beliefs about how the world is basically constituted." However, the role of metaphor in a living mythology is quite significantly different from the role of metaphor in science. Science has its own mythology, but it needn't be a *living* mythology, and when it is used this way it can create severe problems.

Midgley has been consistently lucid on the role of metaphor in science, recognizing (2003) that beyond the data and theories of science there is a "huge, ever-changing imaginative structure of ideas" that scientists use to interpret facts, and that these metaphors and images "cannot possibly be objective and antiseptic" in the way that we usually think about the scientific endeavor. This mythology of science draws from everyday experience, and is intended to relate scientific conclusions to everyday life. Once science has used them, however, "they are often reflected back into everyday life in altered forms, seemingly charged with a new scientific authority." As a result, we have to keep a close eye on the myths of science:

> Metaphorical concepts... are quite properly used by scientists, but they are not just passive pieces of apparatus like thermostats. They are living parts of powerful myths – imaginative patterns that we all take for granted – ongoing dramas inside which we live our lives.... When they are bad they can do a great deal of harm by distorting our selection and slanting our thinking. That is why we need to watch them so carefully.

Again, we can see here my point about the role of fiction in our experiences. We mislead ourselves when we treat fiction as something permanently fenced off from the real world, with no possible connection to it, because fiction is intimately involved in *all* our experiences of the world, including those that come to us through science. When the myths that we rightly use to understand empirical research go beyond their role as explanatory metaphor, they can actively distort our understanding of the world. As Midgley observes, ideological outgrowths of science must be taken seriously because "by claiming its authority, they have injured its image". Consequently, anyone wishing to defend science needs to make the effort to understand the relationship between science and fiction.

Orthodox Science Fiction

The phrase 'hard science fiction' has always troubled me, because it seems to represent a contradiction in terms. Science fiction, by definition, is a form of fantasy, one in which scientific knowledge and technology are the inspiration (as opposed to the magical worlds of sword and sorcery, for instance). Fans of 'hard science fiction' want to read fantasies in which their understanding of science is not transgressed – this is how the phrase is used, and Wittgenstein's advice that the meaning of a word or phrase "is its use in the language" can be gainfully applied here (Wittgenstein, 1962). But in respect of the kind of stories that are told as 'hard sci fi' – often intergalactic adventures, constrained by the limits of special and general relativity, contemporary biology and so forth – we can be almost certain that our *current* understanding of science and technology is radically insufficient to allow us to predict with any confidence just what will be involved in such distant and speculative endeavors as interstellar travel.

Consider how wildly wrong the science fiction of the early atomic age was about what was to come: we did not get our promised flying cars, humanoid robots or nutrition pills. Technology was to move in a very different direction to what was imagined by the writers of the 1950s. Similarly, if mankind is to explore beyond the solar system, it will happen at a time and in a way radically impossible to predict now. Even the colonization of our solar system is difficult to adequately anticipate. Not even something closer to home like moon colonies simplifies our task to the point of it becoming entirely straightforward. We don't have the requisite knowledge or resources to make these fantasies happen, and if and when we do, they will inevitably unfold in a manner that diverges from what we would imagine when thinking about such problems now.

So what exactly is being expected of 'hard science fiction'? I believe it is illuminating to interpret the demand for fantasies that do not transgress an individual's scientific beliefs by considering this strange genre as *orthodox science fiction*. The parallel with religious doctrine may rankle – science, after all, is not a religion, and talking about it as if it were is therefore misrepresentative. However, as previously noted, both science and religion share in common a crucial role for imaginative metaphor, and both function as sources of authority. When we talk about orthodox religion, we often mean those versions of a particular tradition in which the doctrines and mythologies are rigidly constrained to their authorized interpretations. When I talk about orthodox science fiction, the same kind of doctrinal restraint is in effect, but (curiously!) affecting science fiction stories that are supposed to exist purely for entertainment.

The mythologist and historian Charles Segal (1986) coined the term *megatext* to refer to the Greek myths when taken collectively to imply a single fictional world, and this term has been taken up and used in the context of modern science fiction and fantasy franchises. Thus *Star Trek, Star Wars*, Middle Earth, James Bond, Marvel Comics, *Dungeons & Dragons* and so forth each comprise a megatext. These are all works of fiction, but they function in a manner exactly parallel with historical mythologies. They are not, as Campbell (1972) puts it, *living* mythologies, which is to say, they are not mythologies that belong to an extant religion (as, for instance, the stories of the *Ramayana* and *Mahabharata* are for Hindus) but it is chiefly this element which distinguishes the megatext of (say) *Star Trek* from the megatext of Native American mythology.

Science fiction *taken as a whole* also functions as a megatext, as does sword and sorcery when taken as a whole, and for that matter superhero stories. Fantasy author Michael Moorcock often quotes a fellow writer with respect to the way genre fiction functions in practice: "Terry Pratchett wisely said that genre is a big pot from which you take a bit and to which you add a bit." (Brown, 2006). No-one owns the megatext of science fiction, fantasy or horror – they are collectively shared mythologies (although not living mythologies in Campbell's sense). Thus science fiction novels take place within the wider mythology of the science fiction megatext. It is acceptable for these novels to feature starships, faster than light travel, teleportation, humanoid robots or psychic powers because these are all part of the wider mythology, even though specific fictional worlds might reject certain elements. It is clear on this reading that 'hard science fiction' represents a subset of the megatext for science fiction.

It is my suggestion that *science itself functions as a kind of megatext*. For instance, when something is dismissed as "not scientific" it is not usually meant that it does not adhere to the standards of the contemporary paradigms – it usually means that the thing in question is incompatible with the beliefs of people who accept some hypothetical common core of scientific theories and experimental results. So, for instance, the claim that astrology is "not scientific" seems intended to mean that astrology is *false*, not that astrology isn't a scientific research program (it seems unlikely there is *anyone* who thinks astrology is a research program of this kind). This kind of statement isn't even a claim that 'experiments have shown astrology is false' – it is more commonly a presumptive claim that the causal mechanisms deployed in astrological practice are incompatible with something being called 'science'. What is that something?

I suggest that the something in question is what we might call *the science megatext*. There is a collection of things that can be broadly recognized as validated by the current paradigms of science – the theories, the experiments, and the metaphysical beliefs that are assumed to underpin both, such as physicalism (Neurath, 1931). This latter belief constitutes a prior metaphysical commitment shared by the majority of contemporary scientists, as observed by Lewontin (1997):

> It is not that the methods and institutions of science somehow compel us to accept a material explanation of the phenomenal world but, on the contrary, that we are forced by our a priori allegiance to material causes to create an apparatus of investigation and a set of concepts that produce material explanations, no matter how counterintuitive, no matter how mystifying to the uninitiated.

Collectively, these theories, practices and beliefs form a mythology of science, which may include (for instance) beliefs such as 'science evolves towards truth', which as we saw in our discussion of Kuhn is in no way necessary for understanding the practices of scientists. The science megatext is more than the sum of scientific knowledge, since it incorporates an additional, mythological stance concerning what science is, what it will be, what it can be, and perhaps most significantly for our current purposes, what it cannot be.

Thus when a person prefers 'hard science fiction', there is a sense in which they are saying that they want to read stories that are not simply part of the science fiction megatext, but that are also consistent with the *science* megatext. This excludes anything not currently considered plausible by mainstream scientists (such as psychic powers or, for the most part, faster than light travel). The science megatext is thus operating in this context as a kind of doctrine, and the fan of 'hard science fiction' is requiring that the fantasy stories they read that are to fit this term will be orthodox with respect to the current interpretations of the science megatext.

The above claims do not, I will repeat, amount to a claim that science itself is a religion, but rather that the science megatext can function as a mythology, and this is quite apart from its role in guiding scientific research. However, any claim that science has replaced religion, as first suggested by Auguste Comte (1858), begins to treat the science megatext as a living mythology. Positivists in the tradition of Comte effectively claim that *the mythology of science must necessarily replace other mythologies*, and this kind of assertion *does* treat the science megatext as a doctrine. Just as Bertrand Russell (1946) asserted that the political non-religion of Marxism can be considered a religion on account of its inherent mythology, there is an ideology which functions as a scientific non-religion that we can call (after Comte) *positivism*.

Understood in this way, it is unsurprising that we are experiencing an apparent face-off between those who follow traditional belief systems and those positivists with faith in the science megatext, a conflict felt most strongly in arguments over the teaching of evolutionary theories. The flimsiness of the rhetoric on the traditionalist side of this conflict draws attention away from positivist evangelists of the science megatext who attempt to install science in the role of a living mythology. Religion is accused of transgressing the perimeters of science – but has scientific faith, in the form of positivism, also trespassed on the boundaries of religion?

The Borders of Science

Where exactly will we find these boundaries of science that the ignorant hordes of religion are accused of invading? This mythic image of a 'war' between science and religion has become so commonplace that it is all too easy to buy into it, especially in the context of evolution, which serves as the frontline of this alleged battle. This story does not, however, bear up under scrutiny. This does not mean that there isn't a conflict – it's abundantly clear that there is a political fight going on in the United States over the education system, for instance, and echoes of the same can be found in the Islamic world. But as Charles Taylor (2007) notes, the idealized view of 'religion' versus 'science' is an ideological construct that masks an intellectual struggle with complex agendas. As Midgley (2003) observes, the conflict is *political* in nature:

The functions of science and religion within a society are too different for this idea of a competition between them to make much sense once one begins to consider it seriously. Rivalry here only looks plausible when both elements are stated in crude forms (as of course they often are), or when the power-groups that run them conflict at the political level. Political entanglement with power-groups has had a bad effect on religion and does so equally for science, which today is increasingly sucked into the power-struggles of the market.

However, what really complicates this matter is that 'science' and 'religion' are not that easy to apply as clearly defined labels. It may seem that it is trivial to separate religion from its alternatives – especially from the perspective of someone inclined to take a dim view of what religion entails. There's a tendency to presume, for instance, that belief in God or supernatural forces can be taken as a clear marker for religion, but what then of atheistic religions such as Theravada, Ch'an or Zen Buddhism, Taoism, Jainism or Naturalistic Pantheism? A person who insists on equating religion with the supernatural may be inclined to allow these 'godless' belief systems to be excluded from religion, and taken perhaps as 'philosophies'. Yet surely the Human Rights statutes that protect freedom of religion are intended to include these traditions as well, so why make the exception?

The situation is no better with 'science'. Is 'science' intended to mean the knowledge accumulated by experiment and theory? In which case science is in a state of constant flux, and can scarcely make any claim to reliability – just look at the scientists cited in this book who do not agree on key issues in the study of the nature of life. Or is 'science' intended to mean the methods that scientists use in order to gather that knowledge, and construct those experiments? In which case, it is hard to see how 'religion' can threaten it, since individual scientists pursue their research under the framework of their own choosing. Or perhaps 'science' is the systematic endeavor to apply those methods and acquire that knowledge. If so, it wouldn't make sense to talk about conflicts *within* science, yet evolutionary studies are packed full of them – selectionists versus neutralists, form versus function, individualists versus group selectionists and many more disputes besides!

For a war between 'science' and 'religion' to be viable, there must be some common territory. If not, there can be wars *within* 'science', just as there are certainly conflicts *within* 'religion', but there cannot be a dispute between them. The question has to be: is there common ground between the two, and if there is, does it lie beyond Popper's milestone, out in the vast untestable tracts of metaphysics? Because if this is the only place 'science' and 'religion' intersect, we had better learn to live with the problem, because we shall certainly never resolve metaphysical disputes to anyone's satisfaction. As Ruse (2010) has asserted, there will probably always be "some grey or contested areas about the domains of science and religion."

Let us drop the scare quotes for now and pretend that we know what the terms science and religion mean, at least in broad strokes. To some extent, this condemns 'religion' to end up meaning 'Christianity or other religions that are similar to Christianity such as Islam', which I feel is gross caricature of the traditional beliefs of the people on our planet, but alas this cannot be helped. It certainly seems that many of the vocal critics of religion focus their ire on Christians and just assume that all other religions are essentially the same under the hood, but as it happens many of the people who have actively explored the relationship between science and religion have made an explicit choice to treat 'religion' as meaning 'faith tradition' or 'Christianity', if only for practical reasons.

One such intrepid explorer – indeed, the person credited with founding 'Science and Religion' as an area of study – is Ian Barbour. In 1990, he proposed a fourfold typology as a means of sorting out the various ways people have related science and religion and although he has made some minor revisions he had continued to use this system ever since (Barbour 2000). It is widely taught as a means of examining the issues in the interface between science and religion. The four boxes in his model each present a different perspective on how religion and science can or should interrelate.

The first is *conflict*, in which science and religion are seen as enemies. Barbour makes the point that both creationists and atheistic scientists agree on this point, seeing it as impossible for a person to believe in both God and evolution – they only disagree about which to accept. Barbour notes that these two groups "get most attention from the media, since a conflict makes a more exciting news story than the distinctions made by persons between these two extremes who accept both evolution and some form of theism." Richard Dawkins is a prominent example of someone espousing conflict.

The second option is *independence*, in which science and religion are seen as strangers that can get along "as long as they keep a safe distance from each other". This viewpoint denies the validity of any alleged conflict, since science and religion are claimed to refer to differing domains of life. This perspective has a long history – John Henry Newman advocated it in the mid-nineteenth century (Newman, 1873), and it corresponds to geneticist Theodosius Dobzhansky's (1967) remark that "science and religion deal with different aspects of existence" which can be oversimplified to "the aspect of fact and the aspect of meaning." Barbour notes that separating the two fields into "separate watertight compartments" is a way to avoid conflict "but at the price of preventing constructive interaction." Stephen Jay Gould was a prominent example of someone espousing independence.

Thirdly, *dialogue*, in which both similarities and differences are acknowledged, and conversation is facilitated between (say) theologians and scientists. Barbour suggests that dialogue can occur at the limits of science, when it faces a question that it cannot answer such as "Why is the universe orderly and intelligible?" It can also occur when ideas from science are used to influence theological interpretations of the relationship between God and the world. Unlike independence, dialogue doesn't treat science and religion as forever cut off from one another, and unlike conflict, dialogue doesn't recognize a fundamental incompatibility between the practice of science and religious faith. Barbour himself is an example of this approach.

Lastly, *integration*, which is a "more systematic and extensive kind of partnership between science and religion", one that either argues for the reformulation of certain religious beliefs in the light of scientific discoveries (a *theology of nature*, rather than natural theology), or that tries to interpret scientific and religious thought within a common framework (such as process theology). Pierre Teilhard de Chardin is sometimes given as an example of this approach, although Barbour admits to supporting integration as well as dialogue.

Another similar typology has been offered by Haught (1995), one in which the four categories are whimsically given titles beginning with the same letter: *conflict, contrast, contact* and *confirmation*. Barbour recognizes that the first two categories are the same in both systems (i.e. Haught's 'contrast' is the same as Barbour's 'independence'), and suggests that Haught's term 'contact' combines the themes of Barbour's 'dialogue' and 'integration' into one heading – indeed, Barbour notes that "there is no sharp line" separating dialogue from integration, so Haught's conflation in this regard is not unreasonable. The final category, 'confirmation', refers to the validation of scientific thought by background assumptions that originated in theology (such as belief in the rationality and intelligibility of the world), and Barbour suggests that this for him can be considered a form of dialogue.

What can be gained from Barbour's discussions of Haught's closely related typology is that the idea of four categories doesn't quite stack up in either of the two models on offer. According to Barbour, 'integration' is not sharply delineated from 'dialogue', and Haught's 'confirmation' is also a form of dialogue – suggesting that both approaches collapse into just *three* categories – *conflict, independence* and *dialogue* (or conflict, contrast and contact). However, some further conceptual analysis may shed some light onto what exactly it is that either system is supposed to be reflecting.

The essence of the conflict position – whether espoused by a diehard atheist like Dawkins, or the creationists he despises – is a belief in *absolute truth*: there is one account of what is true, and hence once you are sure you have the correct version you can safely dismiss all other accounts as erroneous. This can be contrasted against the independence and dialogue positions, both of which take an attitude of *perspectival truth*. This is not to claim that the truth is inescapably relative or unobtainable (e.g. relativism), but rather to suggest that different perspectives can offer an important part of the true picture and, further, that absolute and objective truth cannot be achieved directly (except, for theists, by God). In the case of independence, the truth is perspectival because science and religion make different kinds of claims about the world; for dialogue, the truth is perspectival because science and religion approach the world from different (but potentially complementary) angles.

If independence and dialogue have this perspectival attitude towards truth in common, what distinguishes them? It is their differing attitude towards the domains (or languages) of science and religion. Independence entails *disjunction*, as both Barbour and Haught assert. On this account, science and religion are different fields, they use different languages, and they talk about different things, as in Gould's proposal of "Non-Overlapping Magisteria" (NOMA) in which science is characterized as "our drive to understand the factual character of nature" and religion is characterized as "our need to define meaning in our lives and a moral basis for our actions" (Gould, 1997, Gould, 1999) – just as in Dobzhansky's simplification of science and religion into aspects of "fact" and "meaning".

Conversely, dialogue (and, by extension Barbour's integration and Haught's confirmation) entails *intersection* between science

and religion; rather than treating the two as if they are partitioned into "watertight compartments", they are allowed to
interrelate. While a non-religious individual such as Gould can
be content with carving up distinct territories for science and
religion, people of faith such as both Barbour and Haught are
generally less comfortable with this solution. This is a point
astutely made by H. Allen Orr (1999) in his review of Gould's
NOMA proposal: "If your religion is dictated by science, the two
are non-overlapping. But if your religion is independent of
science, the two routinely tread on each other's toes." Indeed,
Barbour makes it clear why he believes dialogue is preferable to
independence:

We cannot remain content with science and religion as
unrelated languages if they are languages about the same
world. If we seek a coherent interpretation of all experience,
we cannot avoid the search for a more unified worldview.

In other words, the attempt to compartmentalize science and
religion has the undesired effect for many religiously-minded
people of *demoting* religion with respect to science – fact
somehow outranks meaning, in so much as you want to believe
that both are talking about the same world. There is something
of an echo of the desire for absolute truth that epitomizes the
conflict positions in this approach: if religion has *nothing* to do
with fact, doesn't this come close to saying that religion isn't true
at all?

This conceptual analysis has an additional consequence: the
distinction between disjunction and intersection can *also* be
applied to the conflict camp. As Barbour and Haught have it,
creationists and their atheist opponents all fall into one box. This
is a convenient grouping in so much as it shows up what Taylor
calls the "strangely intra-mural quality" of the alleged war
between science and religion, and certainly in so much as the

atheist faction is conducting their own brand of theology (admittedly, atheology) the disagreements do have some common ground here. But at the same time, there is something deeply distinctive about what Creation Science or Intelligent Design proposes. Their claims rests on a presumed intersection between science and religion – both must conform to a common truth – whereas naturalistic atheism rests on a disjunction – science is true, therefore religion is false.

If my conceptual analysis is accepted, then we are back to four categories in the relationship between religion and science – but they are not quite the same as those proposed by Barbour and Haught. There are two positions based on belief in absolute truth, the absolute disjunction of 'conflict atheism' and the absolute intersection of 'conflict theism'. There are also two positions based on belief in indirect access to truth, the perspectival disjunction of Gould's NOMA and its equivalents and the perspectival intersection of theology of nature and other forms of dialogue. But as Orr suggests, Gould's position begins outside of religion, whereas Barbour, Haught and other advocates of dialogue hold positions that begin *inside* of religion.

In this respect, you might expect an advocate of independence, like Gould, to have some sympathy for conflict atheism – but of course, Gould and Dawkins were bitter rivals (admittedly, mostly over scientific issues). Similarly, you might expect an advocate of dialogue to have some sympathy for conflict theism since they share in common a relationship with religion, but the opposite generally seems true. Barbour says that "creation science is a threat to both religious and scientific freedom", while Haught calls it "a scandal" that there are Christians who accept a literal interpretation of the Garden of Eden and "deplorable that there are still so many defenders of ID and creationism", although he has suggested to me he *does* have some sympathy for creationists even though he opposes their rejection of evolutionary science (Haught and Bateman, 2011). Nonetheless, it seems that whichever camp you fall into, everyone else has it wrong. We can hardly be surprised; it's hard to get anywhere in life if you don't trust in your own judgment.

So who's right and who's wrong? I'm going to take the least popular position and suggest that all four of these camps have an important piece of the puzzle. Both the perspectival positions have it right that absolute truth isn't something we have access to as mere humans, but advocates of disjunction have to massively simplify what religion is allowed to mean in order to make their plan work, and this just isn't acceptable to most people of faith. Conversely, in condemning their more literally-minded cousins, advocates of intersection between religion and science risk denying freedom of belief to those believers unable to make peace with evolution, even though the reason these people are denying evolution has very little to do with science and everything to do with the conflict atheists' clumsy attempts at theology. Both the absolute camps are correctly calling on the faults of their vocal opponents, even though their own positions rest on sand.

Ultimately, if we take seriously the commitment to freedom of belief at the heart of contemporary conceptions of Human Rights, the challenge shouldn't be to establish who is right and who is wrong – as was ever the case, everyone is right in some sense, and everyone is wrong in another sense. The challenge shouldn't be to settle the argument – in so much as key parts of the dispute are far beyond Popper's milestone (e.g. resolving existence claims for God), this simply isn't an option. Besides, we have learned to tolerate conflicts within religion and to accept disagreements inside science as a natural part of the process, so there's no reason to presume that what this issue needs is final adjudication.

What we have to find is not the right answer, but a workable solution that allows everyone concerned to live together. As Ruse (2010) proposes, we need "a great deal of negotiation" in order to clarify the boundaries between science and religion, and even then, Ruse agrees with Barbour and Haught's position of dialogue in suggesting that "science and religion reach across to each other" – they may be distinct, but they are not entirely separate, and "a delicate balancing act is needed". What makes this matter more complicated is that those who have become most emotionally invested in conflict are unlikely to change their minds any time soon. Nonetheless, the challenge of resolving this squabble is important, as Ruse attests, and he is keen to inspire people to get involved. I share his motives in this regard.

In order to begin solving this problem we have to start by setting aside the fiction that the borders of science are to be set against the perimeter of religion. If we continue with the metaphor that all of human knowledge is a territory, we have to recognize that there are kinds of knowledge that are *not* attained by empirical research, such as histories, practical skills (including a vast variety of different crafts), languages, sports and music trivia, not to mention logic and philosophy. Science doesn't have a monopoly on knowledge, and indeed, scientific theories and beliefs change far more frequently than craft skills, trivia or logic. Similar, the religious traditions do not have a monopoly on values and meaning. Musicians, athletes and artisans all have values that are not expressly religious, and so do scientists. The real world is much more complex than these simple fictions allow.

If science isn't the only field with a stake in the metaphorical terrain of knowledge, and religion isn't the only field with a stake in the moral landscape, why are religion and science being painted as contenders? Certainly part of the answer is that the conflict theists do make claims *about scientific subjects* that differ greatly from the official line. However, their motivation in doing so is to oppose *theological* claims made by the conflict atheists that are clearly religious and not scientific in nature. Rather than one battle, perhaps we should see this as two conflicts – and if we do, it ceases to be science versus religion, but (creation) science versus (positivist) science and (atheist) non-religion versus (theist) religion. These are two closely related disputes in political terms, but in terms of what is being fought over, each argument concerns radically separate matters.

We cannot hope to propose a viable treaty between the conflict atheists and the conflict theists – especially since neither side is willing to bend – but neither can we ignore this political battleground, or relegate it to insignificance by calling it the actions of a few misguided souls. I suspect part of the reason Barbour and Haught felt compelled to put both groups into one box was a desire to stress their distance from *either* position, but it is important to treat each as distinct as we have to accommodate both these camps in our plural societies – and each has *very* different needs. If we can understand why each side is so offended by the demands of their 'enemies', we will have taken the first step towards reconciliation, and in both cases this rests on the one thing they have in common: a belief in absolute truth.

Positivism

There can be little doubt that the conflict atheists are positivists in the sense I am using this term, but it would be a mistake to think that they represent the *only* positivists. Taking positivism to mean 'avoiding belief in untestable things' or 'attempting to minimize metaphysical commitments', it becomes clear that many of the people in the independence camp are also positivists. Gould, for instance, or Lewontin, can just as reasonably be characterized as positivists as Dawkins or Dennett.

There are other existing terms that also seem to conform to the concept of positivism – 'scientific materialist', perhaps, or E.O. Wilson's *scientific humanist,* which he stated was "the only worldview compatible with science's growing knowledge of the real world and the laws of nature" (Harvard Magazine, 2005). I suggest we treat all of these labels as denominations or factions within positivism, and treat positivism not as a religion but as a group of non-religious traditions similar to the way the Abrahamic traditions of Judaism, Christianity and Islam can be called 'faith traditions', or the traditions of Hindus, Jains, Sikhs, Buddhists and so forth can be called 'Dharmic traditions'. All these different traditions – religious and non-religious – represent different ways people relate to the world, and all vary in their degrees of dogmatism, the stress placed on individual freedom or on duty to community, and in many other ways beside.

Like a religion, a non-religion can be practiced with differing degrees of vigor. Recall the late-eighteenth century names for "dangerous" religious beliefs: superstition, enthusiasm and fanaticism. The same categories can be applied to the various beliefs collected under positivism. 'Superstition' seems a perfect description for techno-fantasies like the Singularity (what has been called "Rapture of the Nerds"). 'Enthusiasm' applies to those positivists who have attained strident certainty on some-issue-or-another, especially atheology. 'Fanaticism' applies to those willing to act beyond the accepted bounds of polite society, such as the Stalinist regime in the Soviet Union during the early twentieth century. There are, of course, also a lot of moderate positivists, who are embarrassed by such excesses, but moderates are always invisible when your primary sources of information are actively searching for exciting things to report. As Barbour noted, extreme people are a lot more interesting precisely because of their intemperance.

The "New Atheists" (Wolf, 2006) are not, as is sometimes claimed, fanatics, but it is entirely reasonable to consider them *enthusiastic* positivists. Dawkins, Dennett, Ayer and their chums know what they think and they're not shy of sharing their point of view. Really, there are plenty of things being said in the world more shocking than anything this bunch says. In a quirkily self-defeating fashion, however, they ardently believe evolution should be taught to children, yet tell people evolution proves there is no God. It's scarcely surprising that this causes conflict with those people for whom faith in God is an important part of their lives, nor that this consequently makes teaching evolution into an uphill battle.

Neither is this kind of enthusiastic positivism entirely new. T. H. Huxley, known as 'Darwin's bulldog' in the nineteenth century for his advocacy of Charles Darwin's theory of evolution, was part of a political movement in Victorian Britain that challenged the control that the Christian church held on education at that time. Ruse (2003b) observes that Huxley saw the new theories of evolution as just what was needed to break that stranglehold:

> Huxley and his friends needed a new ideology for a new age – they needed their own religion or religion-equivalent to offer the public. What better choice than evolution? It became a kind of secular religion, a modern metaphysics, one that answered questions about origins, as did Christianity; that told us our place in the scheme of things, as did religion; that offered an overall meaning to history, namely, Progress rather than Providence.

Ruse suggests that many positive developments came out of Huxley's efforts, but he does not deny that what was being offered was a substitute for religion – a non-religion to challenge Christianity. As Ruse (2003a) puts it, Huxley "saw the need to found his own church, and evolution was the ideal cornerstone". Indeed, he "aided the founding of new cathedrals of evolution" in the form of natural history museums packed full of fossil dinosaurs, that were being uncovered in large numbers in the United States at the time. The parallels between the architecture of a cathedral and the Victorian museums, Ruse suggests, is not merely coincidental.

As Midgley (1992) puts the matter, Huxley thought of science as "a vast interpretative scheme which could shape the spiritual life, a faith by which people might live." This faith in science served as "a competitor with existing religious faiths, not a way of having no faith at all", and it was this – not scientific disagreements – that seemed to put religion and science into competition. Huxley's evolutionary non-religion promised "a new wisdom, a decisive spiritual and moral advance" (Midgley, 2003). It is debatable that this was achieved.

Ironically, given that the positivists who are today upholding what Huxley began almost uniformly identify as atheists, it was Huxley who coined the term 'agnostic' to describe his position on theology. But then, it is worth noting that nothing in positivism expressly *requires* atheist beliefs; indeed, the case can be made that since positivism is a belief system intimately connected with the values of science, agnosticism would be the more appropriate position. Atheism, almost by definition, leans towards enthusiasm in that it expresses certainty on a clearly untestable issue.

It is important to understand that positivism, while certainly a legitimate belief entitled to the same protections extended to religious and other traditions, is by no means necessary for the practice of science, neither is it inevitable that practicing scientists must choose positivist beliefs. The confusion between science, that is, the methods and practices of empirical research, and positivism, that is, trust in "the power of science to clear up all confusion about the world" (Haught, 1995) has been one of the most damaging things about the debacle surrounding contemporary atheism.

As Haught observes, those who place their trust in the scientific method "cannot scientifically justify this faith without logical circularity". Positivists can make the claim that their belief system requires *less* of a leap of faith than, say, Christianity, but they can't claim that there is no element of faith at all. If a key positivist argument against religious faith is that it is unfalsifiable, then we have to also recognize that positivism is equally unfalsifiable: both positivism and religious faith go far beyond Popper's milestone.

Positivism is thus a different kind of metaphysical faith, as Haught, Midgley, Ruse and many others rightly attest, and this conclusion cannot simply be dismissed as stemming from religious bias: while Haught is a Catholic, Midgley is non-Christian and Ruse is an outright atheist. This is one topic on which everyone (except, perhaps, the more enthusiastic positivists!) agrees. What's more, when positivism masquerades as science, rather than being recognized as a belief system essentially extraneous to scientific practice, it attempts to clothe itself in the authority of science without warrant – and this is the same kind of distortion of metaphysical issues that the Intelligent Design movement and the like are charged with executing. Indeed, it is precisely this failure to recognize the distinction between metaphysics and science that has created such a political impasse over the teaching of evolution in the United States – and the conflict atheists are just as guilty as the conflict theists in this regard!

Although positivism is not a religion as such, it has as much in common with religious traditions as it does with science itself, provided science is taken to mean theory and research, especially if we take seriously the independence thesis advocated by Gould and others that assigns meaning and values to the domain of religion. Haught (2010) makes the point that many of the more vocal positivists ascribe to an "ethic of knowledge", which can only be understood as a moral stance. Dawkins, Dennett and so forth "adhere passionately to the belief that scientific knowing is the only morally right way to put our minds in touch with truth, and they cast scorn on anyone who refuses to accept their exacting standards of ethical existence", and this despite the fact that this claim to the "absolute inviolability" of the ethic of knowledge is presented without any serious philosophical foundations. This problem of grounding morality is by no means unique to positivism, but there is certainly nothing that allows positivists to sidestep it.

Dawkins speaks quite freely about his commitment to the truth (Brockman, 1995):

> I'm considered by some to be a zealot... As far as being a scientist is concerned, my zealotry comes from a deep concern for the truth. I'm extremely hostile towards any sort of obscurantism, pretension. If I think somebody's a fake, if somebody isn't genuinely concerned about what actually is true but is instead doing something for some other motive, if somebody is trying to appear like an intellectual, or trying to appear more profound than he is, or more mysterious than he is, I'm very hostile to that.

This deep concern for the truth is a quintessential positivist value, but it rests on metaphysical commitments. Indeed, I find it difficult to avoid characterizing this belief without suggesting that Dawkins and other positivists consider the truth to be *sacred*. Obviously one meaning of the term 'sacred' implies some connection with a deity or a divine aspect, but it also means "secured against violation or infringement by reverence or a sense of right", as in a "sacred oath" (Random House, 2011). This is a secular sense of the word 'sacred', but it is not easy to find an appropriate synonym that resolves the inevitable tension between the usually antithetical terms 'secular' and 'sacred' in this context.

That truth is seen by positivists as a supreme value is evidenced by the outrage felt by those things that transgress their standards for empirical truth, such as 'Creation Science'. Anyone, such as Dawkins, who believes that the truth is sacred in the sense I have outlined here, i.e. a matter for respect, something that should not be violated or infringed, will naturally feel anger or disgust when the truth seems to have been manipulated or distorted. But this intense commitment to truth isn't a requirement for science – I conduct scientific research into how and why people play games, but I am quite agnostic about truth, and certainly don't see it as sacred. Positivist beliefs may be common among scientists, but positivism *is not science* but rather a non-religious, metaphysical faith in science.

Nietzsche (1887) was perhaps the first to recognize the quasi-religious quality that can creep into the scientific endeavor. In the same book which contains the oft-referenced story of a madman who comes to town asking after God and crying out that "God is dead", he writes about the relationship between truth and science under the heading "How we, too, are still pious". Nietzsche is usually presented in the context of his opposition to Christianity, but his critique aimed at considerably broader targets. Christians will certainly find large parts of Nietzsche's philosophy hard to swallow, but the nineteenth century philosopher has far more to offer than just anti-religious sentiments.

While Nietzsche was adding an explication of his pithy phrase "God is dead", he followed it immediately with a warning to scientists, noting that in genuine science "convictions have no rights of citizenship" – that is, we can only produce provisional beliefs, hypotheses, which only then may earn their status as knowledge through experiment, and even then this status is provisional. Nietzsche shrewdly observed that convictions must cease to be so before they can be science, and thus even before such a process can begin there must be a *prior* conviction "one that is so commanding and unconditional that it sacrifices all other convictions to itself." He explains:

We see that science also rests on a faith, there simply is no science "without presuppositions." The question whether *truth* is needed must not only have been affirmed in advance, but affirmed to such a degree that the principle, the faith, the conviction finds expression: "*Nothing* is needed *more* than truth, and in relation to it everything else has only second-rate value."

This critique of science is often overlooked but it is crucial to a full understanding of Nietzsche's position. The affirmation of the value of truth outlined in Nietzsche shows how truth can operate as a value, indeed, as the highest possible value in the perspective that is presented above. He concludes his analysis of the role of faith in scientific practice by observing:

...it is still a *metaphysical faith* upon which our faith in science rests – that even we seekers after knowledge today, we godless ones and anti-metaphysicians still take *our* fire, too, from the flame lit by a faith that is thousands of years old, that Christian faith which was also the faith of Plato, that God is the truth, that truth is divine...

Nietzsche thus provides an explanation for why positivists often behave as if the truth were sacred. As the scientific endeavor is traditionally pursued, it is underpinned by a metaphysical faith in truth which serves to elevate truth to the status of an ultimate value. As Nietzsche observes, science inherited this faith from Christianity, who borrowed it from Plato. While scientific research itself is not a religion, it originated within religious traditions, and even now that this connection has largely been severed there is still something akin to religious faith that can become invested in science. As Ruse (2003b) puts it, the contemporary conflict over evolution is "a family quarrel – if indeed it is a quarrel at all".

Positivists who have faith in science treat the truth as sacred, and consequently those who violate canonical interpretations of truth are functionally equivalent to blasphemers. Again, what we find here is not a conflict between *science* and religion, but conflicts within metaphysics and ethics between *positivism* and specific religious traditions. As Haught admits, the reason many positivists find religious thought "irrelevant, and even repugnant" is that certain factions within Christianity and other faith traditions have been engaged in "the deliberate avoidance or rejection of evolutionary biology" and this is positivist heresy. But it is positivism, or rather enthusiastic positivism, that creates the conditions for this denial of evolution by mistakenly treating theological issues as scientific and asserting that evolution disproves God. This is a disagreement over what is sacred; the dispute over evolution is a consequence of this deeper conflict, not its cause.

Treating the political battle over intelligent design as purely a matter of science obscures the metaphysical issues at its root, and makes resolution difficult – perhaps impossible – to reach. We need to recognize that wildly different mythologies are colliding: on the one hand, positivists with faith in the science megatext and an orthodox science fiction mythology; on the other, people with faith in God and their traditional religious teachings. As Campbell (1972) knew all too well, Nietzsche saw this "dissolution of horizons" coming, and with the boundaries between belief systems torn down "we have experienced and are experiencing collisions, terrific collisions, not only of people but also of their mythologies."

Positivism is one of many belief systems available today, and in so much as we wish to uphold our notions of freedom it is imperative that it remains *one of many* choices. Commitment to truth as a supreme value, as Nietzsche critiqued, threatens to flatten the experience of life, regardless of whether this "tyranny of truth", as Paul Feyerabend (1988) put the matter, originates in religious mythology or the science megatext. The resolution of the cultural conflict between followers of traditional religious beliefs on the one hand, and enthusiastic positivists on the other, will not come from insisting that any one mythology must be adopted by everyone.

Creation Myths

Mythology has always fascinated me, and for as long as I remember I have been engrossed in the stories that different cultures have told of gods and heroes. I have also always been enthralled by science – to the extent that has sometimes obscured my lifelong engagement with the religious traditions. When the topic of religion came up while I was visiting an old school friend of mine who now lives in San Francisco, he was shocked when I pointed out that as a child I had been a practicing Christian, objecting that I had "always been so into science". The idea that I could be a Christian *and* interested in science was apparently hard to accept!

People who lean towards positivism have no problem accepting the role of mythology in religion, but they find it much harder to recognize the role of myths in science. Katherine Hayles (1995) observes that "evolution is a science" but "it is also a narrative" and no narrative exists in isolation – it is always part of a wider pattern i.e. Hayles' grand narratives or Campbell's living mythologies. As Midgley (2003) recognizes, we tend to see myths as the exact opposite of science, yet they are rather central to science, being "the part that decides its significance in our lives". Part of the problem is the common usage of the term 'myth' to mean a lie or a fabrication, but while myths are always fiction, this does not preclude them from conveying truth – this is precisely what I mean by truth from fiction.

Throughout this book I have talked about 'the nature of life', but even this is a myth, an imaginative metaphor, and precisely the difficulty in ascertaining the answer to the question 'what is the nature of life?' is that both life and nature are concepts that approach infinity in scope and complexity. There is no single or clear answer to this question, neither will one emerge over time. The nature of life is inherently mysterious, and attempts to chart the mechanics of the living world seem to void this inscrutability only when metaphor-laced ideologies substitute a caricature of life and nature for its intricate and multifaceted reality. Empirical study can produce fascinating perspectives on tiny aspects of the nature of life, and theoretical frameworks can attempt to tie these together into bolder claims – but in so doing, mythologies are inevitably formed or developed. Grand theories require grand metaphors, and these will always tend to be mythic in their approach.

Midgley and I use 'myth' to mean "imaginative patterns, networks of powerful symbols that suggest particular ways of interpreting the world", and myths, in this sense, deeply affect the way we experience the world and the meaning we find in it, and are an entirely inescapable part of who we are: "We have a choice of what myths, what visions we will use to help us understand the physical world. We do not have a choice of understanding it without using any myths or visions at all." (Midgley 1992). This is another way of saying that our fictions are an intimate and inescapable part of who we are. If we seek the truth, we will not find it by avoiding fiction but only by understanding our relationship with it.

Evolution is "the creation myth of our age" but again this *does not* mean that it is false, but rather that "it has great symbolic power, which is independent of its truth" (Midgley, 1985). The creation myth of evolution, however, is not a part of science, even though it is shaped by scientific research. There is, as Ruse (2003a) makes clear, a distinction between evolutionary science and "evolution as a secular religion" – even if the same people sometimes talk about both subjects without clear distinction. Ruse doesn't claim that this "is all bad or that it should be stamped out", but he does not deny that "popular evolutionism" functions as an alternative to religion.

If we accept the story of evolution, it forms a part of our perspective on the world, and this cannot help but change the way we look at the universe around us. Indeed, this transformation of perspective is one of the reasons that 'popular evolutionism' can function as an alternative to religion, albeit a kind of positivist non-religion with almost no stake in traditional mythologies. Once the implications of evolutionary science are accepted, the creation myth of evolution becomes a part of our own story about the world, but how we incorporate the myth of evolution into our other mythologies is something that every individual has to work out for themselves.

There will be some for whom the evolutionary creation myth cannot be accepted. We may see this as unfortunate, but if we uphold freedom of belief we have to allow people the choice of what to believe – even if what they believe substantially contradicts the science megatext. Equally, there will be those for whom the myth of evolution will render the older mythologies unworkable. Edward O. Wilson (1998) suggested that "the true evolutionary epic, retold as poetry, is as intrinsically ennobling as any religious epic" and that the findings of science "already possesses more content and grandeur than all religious cosmologies combined." I don't agree, but then, I'm not a positivist. For me, as amazing and exhilarating as I find the evolutionary science and mythology, the sacred narratives of the world have so much more to offer.

It's ironic, but not entirely unexpected, that the evolutionary creation myth has come into conflict with traditional Christian (and Muslim) views of creation. The irony comes from the fact that if it were not for the Biblical perspective on time, with a definite beginning and end, contemporary evolutionary theory might never have come about (Haught, 1995). As Ruse (2003b) puts it: "It took the ancient Jews to make our world a history, to make questions about ultimate origins meaningful, along with future expectations and consequences." The faith traditions that trace their ancestry to Abraham were intimately involved in the circumstances that made it possible for an evolutionary myth of any kind to come about. Under the influence of Eastern mythologies, which have always preferred to view time as cyclic, it's not clear how what we now call evolution could have come to light.

The story of Darwin's big idea is often presented as a cataclysmic break with traditional religion – but this is a myth in every sense of the word. In his account of the changes in belief over the last five hundred years, Charles Taylor observes that Darwin's theory "didn't emerge in a world where almost everyone still took the Bible story simply and literally". However, Taylor accedes that "science (and even more 'science') has had an important place in the story". The universe revealed by evolutionary theory was wildly different from the hierarchical cosmos that "our civilization grew up within" and the new perspective made it difficult to see how humanity could have "any kind of special place in the story". There is thus a certain perspective from which "'Darwin' has indeed, 'refuted the Bible'."

It is an open question, however, whether the interpretation of Genesis which 'Darwin refuted' is a necessary component of Christian faith. Haught (2010) believes "many, if not most" Christian theologians accept evolution, and there are a great many Christians in the United States today for whom the idea of a conflict between evolutionary theory and religious belief makes little sense. As the National Academy of Sciences has reported, the religious denominations which do not accept evolution "tend to be those that believe in strictly literal interpretations of religious texts." (National Academy of Sciences, 2008). There is also the question of whether Biblical inerrancy constitutes a form of idolatry and is thus inconsistent with Christian teachings – what is sometimes termed "bibliolatry" (Geisler et al, 1980) – but this issue need not be considered in any detail here.

Whatever is to be made of the Christianity of those who reject evolution, *they have every right to choose what they believe.* The issue of evolution in the United States doesn't usually focus on the right for people *not* to believe in it, but perhaps it should. No-one is compelled to believe in any aspect of scientific research, theory or mythology, and this is just as well – as we have clearly seen, scientists don't fully agree on all manner of points, and for a great variety of reasons. There is no easy way of adjudicating disputes concerning the science megatext. There are, however, many aspects of the science megatext that are not in significant dispute – optics, for instance, has ceased to be a significant research subject and has now become a matter of engineering (Kuhn, 1962). It's clear, however, that evolutionary theories are *not* one of these consensus areas. If there's anything in evolution that a broad majority of scientists agree upon, it is Darwin's imaginative metaphor of fitness to environment, but this is principally a heuristically useful fiction.

This begs the question: why *must* evolution be taught in high schools? It is not enough to say that it's true, since there are an unlimited number of truths that are not in consideration for school curricula. No-one is suggesting Einstein's theory of general relativity is a suitable high school topic, for instance, but that doesn't mean anyone significantly disputes its claims. Some subjects make more sense to be taught at the university level, when students can get far more detail on the intricacies of their workings, and it is worth seriously considering whether evolution might not be one of these.

The positivists and their allies who want evolution taught in high schools in the United States are outraged at attempts to exclude evolution from teenage education, and call the Intelligent Design movement a "threat to science education" (Boston, 2005), but I've not heard anyone make a coherent argument for *why* evolution must be taught. The case *against* allowing intelligent design to be taught in science lessons isn't a case *for* the teaching of evolution. It is simply assumed that evolutionary theories should be taught. In as much as a case is made, it is advanced on the grounds that science is not democratic and therefore should not be subject to democratic influence i.e. that the content of science education is to be determined solely by scientists.

The famous report by Lawrence Lerner that graded various US States as to their treatment of biological evolution is an example of this approach (Lerner, 2000). It's true, as Lerner says, that science is not democratic. But then, neither is it autocratic. Either way, decisions of what elements of the science megatext are to be taught are neither straightforward nor the sole purview of scientists. Lerner argues that public schools have "a duty to explain... the consensus of scientists on any particular issues", but how Lerner's consensus is to be ascertained is vague at best. He claims that those with scientific credentials who disagree on evolution "do not contribute to the progress that is the hallmark of science", which is a strangely ideological reason for excluding them from the criteria of consensus. It's a rather convenient concept of 'consensus' that allows anyone who dissents to be ignored since "their arguments are useless".

The kind of argument mounted by Lerner is at least as tenuous as anything Creation Science or Intelligent Design has proposed. He objects that not teaching evolution "harms the teaching of the life sciences" and "makes it difficult for the student to come to a clear understanding of how science works". Regarding the former, I wonder how many life science jobs there are that depend on a grasp of evolution for their practice – none that don't presuppose university study, I'll wager. As for obfuscating how science works, there can be few greater misrepresentations than to suggest that scientific consensus exists at this time on anything but the vaguest general statements concerning evolution.

What I find particularly disturbing in Lerner's rhetoric is the idea that it is vital to science education that the only creation myths that are acceptable are those that accord with positivism. In addition to Christian groups, he appears to find offensive Native Americans "whose arguments begin and end with the assertion that their ancestors have lived in the same place for an infinite time". As an enthusiastic positivist, Lerner apparently cannot understand why anyone would believe anything other than the latest scientific estimate of the age of the world, a number that has changed significantly over the last two centuries.

Lerner does not seem to appreciate that by placing their relationship with the world in the context of infinite time, Native American creation myths (or perhaps, non-creation myths) engender an attitude of stewardship towards the natural world that have helped these cultures survive for some fifty thousand years or more. This traditional sacred truth is incompatible with the positivist's empirical sacred truth but the disagreement centers upon conflicting mythologies and has very little do so with the scientific endeavor. (I doubt any of the Native Americans concerned are teaching geology). It seems almost as if Lerner, although he does not use these terms, believes these 'ignorant savages' need to be civilized into accepting contemporary mythology. This chilling echo of the imperialistic European *mission civilisatrice* that drove colonialism is a stark reminder that positivism is a direct descendent of Christianity, and can be prone to the same excesses.

What's more, it is a fundamental human right that parents get to choose how their children are educated. The raising of children is a matter for families, not for governments, and the government's role should be to support parents and communities in the challenging art of educating the young, not to dictate some systematic doctrine that must be taught to all children. To be honest, this kind of one-size-fits-all approach to education doesn't look very plausible even within one school, let alone across an entire continent. As Ivan Illich (1971) recognized, the education system in the West developed under the auspices of a Christianity that was deeply institutional, creating programs geared towards inculcation of dogma rather than assisting individual learning. We can and should do better when what is at stake is the education of our children.

As far as the United States is concerned, there is another reason that might make teaching evolution in schools problematic. Since enthusiastic positivists like Dawkins present theological beliefs as if they were science, issues of the separation of Church and State enshrined in the first amendment *might* apply to the teaching of evolutionary theories in public schools in so much as favoring one creation myth over another in education serves to prohibit the free exercise of religion. This would only be an issue if atheology was being taught in a science classroom under the guise of presenting it as evolutionary science, but it's not unreasonable for parents to be concerned about this possibility – especially since certain prominent scientists like Dawkins have routinely confused theology and science in their attempts to 'popularize science', or rather, evangelize positivism. This kind of philosophical confusion causes far more 'harm' to the life sciences than not teaching evolution – indeed, it is the principle reason that resistance to evolution is so vehement.

Neither is this problem constrained to the United States. The Jordanian molecular biologist, Rana Dajani, has observed that resistance to evolutionary theories in the Muslim world is rooted in a perceived need to oppose the kind of atheological arguments offered under the veneer of science. Dajani, herself a Muslim, decries the alleged conflicts between Islamic and scientific beliefs, stating that her religious tradition has lost sight of its traditional emphasis on original thinking and the capacity to exercise personal judgment. She also suggests that this ideological conflict in the West lends power to dictatorships in the Middle East, where a lack of freedom of thought "facilitates ignorance about these kinds of scientific issues" (Dajani and Bateman, 2011). Resolving the Intelligent Design debacle is thus important on an international scale as well as internally to the US.

I can understand why positivists and others who place considerable value on the perspectives offered by science feel it is important to teach evolution – once you begin to fathom the awesome implications of what has followed from Darwin and Wallace's work, it does change your perspective on the nature of life. But much of the transformation is to your mythology of the natural world; it's hardly a theory with everyday practical implications. In fact, it's hardly a theory at all, at least not a *single* theory; more a fascinating research domain with many contrasting theoretical frameworks. Studied in depth, as can be done at university, it's an important part of human knowledge. Studied in passing, it risks being nothing more than a contemporary creation myth, something far from essential to a good science education.

I am not a United States citizen, but my son is. At the moment, he's only a few months old, and I have no idea which side of the Atlantic he will be educated, but if he should end up going to school in the US I certainly won't be worrying about whether he learns about evolution in school or outside of it. Most likely he will learn about the nature of life from me, and I will teach him not only the latest concepts that science offers on the origins and development of life but also the many traditional creation myths, each of which has its own unique perspective to offer. My son may grow up to be a positivist, and if he does I will support that. But I will certainly want him to experience more than just one mythology on his way to adulthood.

What I find most distressing about the fight over creation myths in the United States is the sheer scale of the political capital being expended on this issue. The more belligerent positivists must feel a deep piety for the sacred truths of science to believe that this is the most pressing matter at hand, and I wonder what it is that makes their Christian opponents believe that what Jesus most wants from them is a defense of old stories against new ones. When I think of how far that great nation has fallen from the ideals that motivated it to help draft the so-called Universal Declaration of Human Rights in 1948 it appalls me that this petty squabble over high school education is felt to require more urgent attention than blatant human rights abuses that have been brazenly conducted with government assent. To bring shame on the United States armed forces in this way is a profound betrayal of all the brave men and women who fought for the ideals of justice and freedom for all, including my wife's father and grandfather.

Facing Silence

Far more distressing to me than these myopic political tantrums over creation myths, however, is the idea that my son or his children might grow up in a world where they could not sit in their back garden and watch the swifts or the barn swallows arriving for their summer vacations. Since 1994, the numbers of swifts has fallen by more than forty percent, and they are not the only birds whose numbers are in severe decline (Tennekes, 2010). The problem is that the insects on which swifts and many others birds feed are in severe decline, in part because insecticides used in agriculture have devastated insect populations, in part because of a global reduction in biodiversity. The much-reported decline of bee populations is a symptom of the same problem (Kluser et al, 2010).

Neither is this an entirely new predicament – we faced something similar in the 1960s with the pesticide DDT, which led to Rachel Carson's (1962) famous book *Silent Spring*. The words she wrote fifty years ago remain apt:

These sprays, dusts, and aerosols are now applied almost universally to farms, gardens, forests, and homes – nonselective chemicals that have the power to kill every insect, the "good" and the "bad", to still the song of birds and the leaping of fish in the streams, to coat the leaves with a deadly film, and to linger on in soil – all this, though the intended target may be only a few weeds or insects. Can anyone believe it is possible to lay down such a barrage of poisons on the surface of the earth without making it unfit for all life?

Today, the problem is far more complex, involving subtle yet serious disruptions to ecology even when the poisons used target only specific creatures, but the essential crisis remains the same. While we argue amongst ourselves about our creation myths and our sacred truths, the natural world around us suffers a disastrous collapse in diversity. We are not only the cause of this catastrophe, we are also its future victims. As pollinators, bees are vital to the growth of the plants that feed us, or that feed the livestock we also feed upon, and we need the bees a great deal more than they need us. The same is true of a great many species, such as the worms essential to maintaining soil, and the trees that recycle the air we breathe. Our gluttony and disregard for the environment may eventually lead to our extinction; it has certainly *already* led to the extinction of a great many species that shared the planet with us.

Yet despite a widely acknowledged urgency on this issue, there is still a disturbing tendency for the problem to be approached from a perspective of *blame*. Religious people blame positivists for emptying the world of every sacred concept except truth, while positivists blame religion for creating the conditions of the problem in the first place. John Passmore (1974) epitomizes this latter attitude:

Only if men see themselves... for what they are, quite alone, with no one to help them except their fellow-men, products of natural processes which are wholly indifferent to their survival, will they face their ecological problems in their full implications. Not by the extension, but by the total rejection, of the concept of the sacred will they move toward that somber realization.

In other words, Passmore suggests that we should be focused on evangelizing positivism to dismantle all notions of the sacred (except, perhaps, that of the truth as sacred), on the belief that once religion is removed the solution will be straightforward. The fact that positivists are just as diverse in their beliefs as people of faith just doesn't seem to have occurred to him, obsessed as he is in his enthusiastic positivist crusade. Would it not make more sense to try to build a consensus of diverse individuals with a common concern for the environment in order to act in its defense, rather than trying to convert everyone to non-religion? It would certainly be more likely to succeed. Not to mention that suggesting there are no religious people with a deep reverence for nature is an insult to Native Americans, Hindus, Jains, Zoroastrians, Buddhists, Wiccans, and many more individuals beside.

In terms of the ecological 'blame game', Lynn Townsend White, Jr.'s (1964) essay on the roots of the ecological crisis has been influential. White observed how the Bible asserts man's dominion over nature and the inherent inferiority of animals when compared to humanity, and this is sometimes taken as a knock down argument against religion. Less often is it pointed out that White concluded that "more science and more technology are not going to get us out of the present ecologic crisis until we find a new religion, or rethink our old one." In fact, he suggested a mythological solution –St. Francis of Assisi's imaginative ideal of a "democracy" of creation:

...the present increasing disruption of the global environment is the product of a dynamic technology and science which were originating in the Western medieval world against which Saint Francis was rebelling in so original a way. Their growth cannot be understood historically apart from distinctive attitudes toward nature which are deeply grounded in Christian dogma. The fact that most people do not think of these attitudes as Christian is irrelevant. Both our present science and our present technology are so tinctured with orthodox Christian arrogance toward nature that no solution for our ecologic crisis can be expected from them alone. Since the roots of our trouble are so largely religious, the remedy must also be essentially religious, whether we call it that or not. We must rethink and refeel our nature and destiny...

The point made here is important: although White recognized that the roots of the crisis lay in Christian beliefs, the *same* beliefs are held today by many positivists, who inherit them because our scientific tradition descends from Christianity. The problematic attitude is a part of the wider culture, and is by no means something constrained to religion. In fact, scientific myths such as gene supremacy can contribute to this kind of blindness to nature – the idea that we are robots programmed by our genes is hardly the mythology we need to build respect for the natural world. It rather tends to have the opposite effect, eliminating any need to respect animals or nature since all the tapestry of life *really* amounts to is the expression of gene patterns. We need new mythologies in order to reinvent our nature and our destiny, as White suggested.

Haught (1995) is a theologian who is working on the challenging, but not insurmountable, task of rethinking religious mythology, recognizing that the environmental crisis creates a sense of urgency compelling people of differing perspectives to arrive at a shared concern for the natural world. He sees our current situation as "one of those special historical moments when our religion is invited to undergo a radical transformation... to bring to it a new vitality." He states that we cannot expect the existing religious texts to have ready-made solutions to this problem, because it is so unprecedented. He writes that the cause of the problem is that we have not accepted "the finitude of nature", and consequently have developed a "distressing compulsion to squeeze infinity out of that which is finite."

On the other side of the coin, positivists like Lynn Margulis (1998) propose scientific mythology that is compatible with the needs of our current situation, shifting the focus of attention to our relationship with the environment. It is no wonder her work appeals so widely to environmentalists of many different stripes, since her myths are perfectly suited to the problems we are facing in our ever-worsening relationship with the natural world. Her imagery is fatalistic yet oddly optimistic:

Gaia, in all her symbiogenetic glory, is inherently expansive, subtle, aesthetic, ancient, and exquisitely resilient. No planetoid collisions or nuclear explosions have ever threatened Gaia as a whole. So far the only way in which we humans prove our dominance is by expansion. We remain brazen, crass, and recent, even as we become more numerous. Our toughness is a delusion... We people are just like our planetmates. We cannot put an end to nature; we can only pose a threat to ourselves. The notion that we can destroy all life, including bacteria thriving in the water tanks of nuclear power plants or boiling hot vents, is ludicrous. I hear our nonhuman brethren snickering: "Got along without you before I met you, gonna get along without you now," they sing about us in harmony.

These new mythologies are a step towards a solution, but they are not the whole answer. Environmentalism is another non-religion equally prone to the excesses of 'dangerous beliefs' –

whether it is expressed in the superstition of various New Age cults, the enthusiasm that denies any need for discussion concerning how we should live with nature, or the fanaticism of animal rights activists who commit violence against human animals in the defense of their non-humans relatives. Those who campaign for the environment are welcome to come to the table and provoke some vigorous debate, but they cannot expect that the problems can be solved without discussion.

Charles Taylor offers a particularly pertinent myth for understanding our current situation, in terms of the incalculable diversity of belief in the world today. It is no longer sufficient to think of the belief systems of the world solely in terms of traditional religion, nor in terms of 'religion' versus 'science'. The decisive breaks with Christianity in the preceding centuries set up two polar extremes – orthodox traditional religion at one pole, and orthodox positivism at the other. But most of us are not at these extremes, but somewhere in between, in a "steadily widening gamut of new positions – some believing, some unbelieving, some hard to classify". He calls this explosion of beliefs the *nova effect*, stating we are "living in a spiritual supernova, a kind of galloping pluralism on the spiritual plane." The myth of the nova effect draws attention to the hopelessness of trying to solve our problems by advocating conversion to one single belief, since this option has long since vanished: "Our world is ideologically fragmented, and the range of positions are growing as the nova effect is multiplied by expressive individualism."

This fragmentation of belief does not mean traditional religion no longer has a voice in the world – many people who are not followers of a specific religious path still listen to, and are willing to be influenced by, those that do. What the nova effect does mean, however, is that traditional religion must share its platform with positivists and other non-religious individuals. Mythologies, and the ethical ideals they foster, have collided – as

Nietzsche predicted – and the aftermath of that titanic explosion is the nova effect, the near-infinite diversity of beliefs we now find in our world.

It is for this reason that the solution to any and all of the problems we face, including that of our relationship with the natural world, must begin by acknowledging the essential diversity of opinion and belief. The old imperialistic solutions that presumed one race or people had a monopoly on the right answers led to the disastrous collapse of those empires in the twentieth century. Positivists hoping that science can pick up the baton of empire and lead us to a bright new future should reconsider their motives and goals. Whatever it is that we want to achieve, we shall certainly need to discuss it further. Insisting on absolute truth is to relegate oneself to the margins – the media may get a guilty thrill in publicizing this kind of pugnacious ignorance, but we need to take seriously the implications of perspectival truth.

Immanuel Kant (1785) had his own mythological image of how we might live: a *realm of ends*, the state of communal autonomy where, despite having different goals and ideals, we find some way to work out our disputes and harmonize our objectives. He admitted this would not be easy, in fact, he said that it was "merely possible" that it might be achieved, yet this mere possibility was enough to motivate the attempt. Ever since reading my first book of Kant's ethics, I have found this myth to be utterly engaging – I believe it touches the heart of what Jesus called "the Kingdom of Heaven", what John Lennon called "a brotherhood of man", what Martin Luther King Jr. called "the promised land", and what many more sages and heroes have strived for besides.

In the wake of the nova, we cannot expect to attain consensus as easily as our cultures did when the horizons of their world were narrow and parochial. Agreement is easy when everyone believes similarly, it is far harder to attain when everyone thinks differently. Yet the presence of diverse points of view can make it *easier* to find the truth, not harder, since we can now see the world from so many unique perspectives that the bigger picture can be even greater and more complete than it ever was before. What has become harder is to convince people that *your* truth is the *only* truth – and this is good news for everyone, because this kind of absolute truth is at best a stick we can use to beat up other people who think differently. Even those who believe they are paragons of liberal acceptance end up demonizing those who are not so broadminded – liberal tolerance falls into the easy rut of absolute truth just as much as traditional conservative politics.

Working towards the kind of fractious consensus implied by combining Kant's myth of a realm of ends with Taylor's myth of the nova effect means a reversal of how we normally think about solving problems. There is a tendency to think in terms of *finding* the solution, and *then* implementing it. But communal autonomy means that finding the solution and implementing it are part of the same process – that of having the discussions that establish what kind of options are available, and choosing which to pursue. This may not mean *debate*, per se, since seven billion people are scarcely in a position a mount this kind of formal discussion, but it has to involve a network of communications. I would scarcely be proposing something new if I gestured to the internet as the only forum capable of housing this global village meeting, nor would I sound too radical if I suggest that leaving matters such as these to the politicians is not a plan for success.

The alternative mythology of evolution I have developed in this book was drawn from the work of scientists, but it is ultimately compatible with our religious traditions. The *chain of inheritance* serves to reposition questions of destiny in terms of our relationship with the past, instead of the future – a kind of thinking integral to the Jewish religion, which sees itself as an unbroken chain. *Refinement of possibilities* relates to contemporary Christian theological ideas, which view the future as a promise to be fulfilled. *Advantages persist* can be read within any religious conception of blessings. Similarly, *trust is an advantage* fits with any number of religious perspectives, especially Islamic notions of hospitality and Christian concepts of fellowship. The *metaphor of design* is compatible with any notion of Creation read symbolically. The *conditions for existence* relate to certain Eastern religious concepts, including Dharma if it is understood as a boundary of a person's viable options for life, or the Hindu concept of samsara as a perpetual flow of beings. *Truth from fiction* is embedded in every traditional folk story with ethical overtones.

I did not offer this alternative mythology in order to reconcile evolution with our religious traditions, since I do not believe that the perennial wisdom can possibly be undermined by empirical theories. But I do offer this mythology as one that could be accepted by both positivists and traditional religious practitioners alike. It is no longer reasonable to accept a cultural conflict between positivist non-religions and the traditional Christianity they descended from: since the defenders of the old ways will not bend, their opponents have an opportunity to decide if they are equally inflexible and dogmatic or, alternatively, if they are capable of focusing on the battles that really matter. Evolutionary theories offer a new perspective on the history of the world, one in which the potential for extinction hangs above every species that runs amok and exhausts its own habitat. If we find just one lesson for today buried in nature's past, let this be it.

However, there is a tendency to assert that the environmental crisis is so urgent that we will all be dead tomorrow – this kind of hysterical panic isn't helpful. We have caused a great deal of damage to the world we live in, and already lost forever many species that were travelling with us until only recently, but nothing suggests that our doom is imminent. The only inevitability is that if we change nothing in the way we live, we won't be around much longer. In the light of the challenges we face, choosing myths that convey a sense of helplessness, or that dismiss animal life as secondary to inheritance mechanisms, is scarcely a wise way of imagining our situation, and clinging to arrogant myths of absolute truth while denying others the right to their beliefs is tantamount to suicide.

We all face silence eventually, as individuals and as species, even if some believe that we may echo in eternity. But the chain of inheritance that connects all life goes on. Our oldest ancestor appeared on this planet some four and half billion years ago. We are about halfway through the life of our star, the sun, which has perhaps another five billion years to go. When I marvel at the mythology of evolution, I wonder whether any of my descendents, or the descendents of animals like the swifts that share this world with me, will be around to witness the end of the party. With all the advantages that we have inherited, with the incredible refinement of possibilities that has occurred thus far, it's at least possible that history is only just beginning.

References

Alexander, Richard (1987). *The Biology of Moral Systems*, New York, NY, Aldine De Gruyter.

Ariew, André and Lewontin, Richard C. (2004). 'The Confusions of Fitness', *British Journal for the Philosophy of Science*, vol. 55, pp. 347-363.

Ariew, André and Ernst, Zachary (2009). 'What Fitness Can't Be', *Erkenntnis*, vol. 71, no. 3, pp. 289-301.

Aristotle (circa 350 BC). *Metaphysics*.

Arnold, A. Elizabeth; Mejía, Luis Carlos; Kyllo, Damond; Rojas, Enith I.; Maynard, Zuleyka; Robbins, Nancy and Herre, Edward Allen (2003). 'Fungal endophytes limit pathogen damage in a tropical tree', *Proceedings of the National Academy of Sciences of the United States of America*, vol. 100, no. 26 (December), pp. 15649-15654.

Ashby, Ross (1956). *An Introduction to Cybernetics*, London, Chapman and Hall.

Associated Press (2006). 'Hamster, snake best friends at Tokyo zoo', 24 January [online] http://www.msnbc.msn.com/id /10903211/ns/world_news-weird_news/t/hamster-snake-best-friends-tokyo-zoo/ (Accessed 9 August 2011).

Astill, James (2002). 'Lioness adopts another antelope', *The Observer*, 17 Februrary [online] http://www.guardian.co.uk/world/2002/feb/17/jamesastill.theobserver (Accessed 9 August 2011).

Avery, Oswald, MacLeod, Colin, McCarty, Maclyn (1944). 'Studies on the chemical nature of the substance inducing transformation of pneumococcal types. Inductions of transformation by a desoxyribonucleic acid fraction isolated from pneumococcus type III'. *Journal of Experimental Medicine*, vol. 79, no. 2, pp. 137–158.

Badiou, Alain (2001). *Ethics: An Essay on the Understanding of Evil*, trans. Peter Hallward, New York, NY, Verso.

Barah, David P. (2009). 'We Are All Madoffs: Our relationship to the natural world is a Ponzi scheme', *The Chronicle of Higher Education*, 31 August 2009, [online] http://chronicle.com/article/We-Are-All-Madoffs/48182/?sid=at (Accessed 3rd February 2011).

Barash, David (1980). *Sociobiology, the Whisperings Within*, London, Souvenir Press.

Barbour, Ian G. (2000). *When Science Meets Religion*, New York, NY, HarperCollins.

Barmore, Laura (2008). 'History of the Lab', 6th August 2008 [online] http://www.alllabs.com/labrador_retriever_history.htm (Accessed 6th August 2011).

Bateman, Chris (2011). *Imaginary Games*, Winchester and Washington, Zero Books.

Baumgartner, Thomas; Heinrichs, Markus; Vonlanthen, Aline; Fischbacher, Urs and Fehr, Ernst (2008). 'Oxytocin Shapes the Neural Circuitry of Trust and Trust Adaptation in Humans', *Neuron*, vol. 58, no. 4 (May), pp. 639-650.

Beatty, John and Finsen, Susan (1998). 'Rethinking the propensity interpretation: A peek inside pandora's box' in Ruse, Michael (ed.), *What the Philosophy of Biology Is: Essays Dedicated to David Hull*, Dordrecht, Netherlands, Kluwer Academic Publishers, pp 17–30.

Bekoff, Marc and Pierce, Jessica (2009). *Wild Justice: The Moral Lives of Animals*, Chicago, Chicago University Press.

Bermudes, David and Margulis, Lynn (1987). 'Symbiont acquisition as neoseme: Origin of species and higher taxa', *Symbiosis*, vol. 4, pp. 379-397.

Booth, A., Shelley, G., Mazur, A., Tharp, G., and Kittok, R. (1989). 'Testosterone, and winning and losing in human competition', *Hormones and Behavior*, vol. 23, pp. 556-571.

Bowlin, Melissa S.; Bisson, Isabelle-Anne; Shamoun-Baranes, Judy; Reichard, Jonathan D.; Sapir, Nir; Marra, Peter P.; Kunz,

Thomas H.; Wilcove, David S.; Hedenstro, Anders; Guglielmo, Christopher G.; Åkesson, Susanne; Ramenofsky, Marilyn; Wikelski, Martin (2010). 'Grand Challenges in Migration Biology', *Integrative and Comparative Biology*, pp. 1–19

Boston, Rob (2005). "Biohazard: 'intelligent design' poses threat to science education and church-state separation, say parents and experts at Pennsylvania trial", *Church and State*, vol. 58, no. 10 (November), p4.

Brock, Thomas D. (1990). *The Emergence of Bacterial Genetics*, Cold Spring Harbor, NY, Cold Spring Harbor Laboratory Press.

Brockman, John (1995). *The Third Culture: Beyond the Scientific Revolution*, Simon and Schuster.

Brown, Helen (2006). 'The heroine addicts' [online] http://www.telegraph.co.uk/culture/books/3654850/The-heroine-addicts.html (Accessed 24 June 2011).

Burgess, Joanna (1999). 'The Last Word', *New Scientist*, vol. 2182 (17 April).

Butler, Rhett A (2006). 'Tropical Rainforests: The Canopy', *Mongabay.com: A Place Out of Time: Tropical Rainforests and the Perils They Face*, 9 January 2006 [online] http://www.mongabay.com/(see browser address bar for URL) (Accessed 3rd August 2011).

Campbell, Joseph (1972). *Myths to Live By*, New York, NY, Bantam.

Campbell, Joseph (2001). *Thou Art That*, San Francisco, CA, New World Library.

Canfield, Donald E., Glazer, Alexander, N., Falkowski, Paul G. (2010). 'The Evolution and Future of Earth's Nitrogen Cycle', *Science*, vol. 330, no. 6001 (October), pp. 192-196.

Carnap, Rudolf (1928). *Der Logische Aufbau der Welt*, translation George, Rolf A. (2003) Peru, IL, Carus Publishing Company.

Carson, Rachel (1962). *Silent Spring*, Boston, MA, Houghton Mifflin.

Comte, Auguste (1830). 'Course of Positive Philosophy', reprinted in Lenzer, Gertrud (ed.), *Auguste Comte and Positivism: The Essential Writings*, New York, NY, Harper, pp. 71-86.

Comte, Auguste (1858). *The Catechism of Positive Religion*, translated by Congreve, Richard (2004), Whitefish, Montana, Kessinger Publishing.

Conway Morris, Simon Conway Morris (2003). *Life's Solution: Inevitable Humans in a Lonely Universe*, New York, NY, Cambridge University Press.

Cuvier, Georges (1800), *Leçons d'anatomie compare*, Paris, Baudouin.

Dajani, Rana and Bateman, Chris (2011). 'Dajani on Evolution in the Muslim World', 21 June 2011 [online] http://onlyagame .typepad.com/only_a_game/2011/06/dajani-on-evolution-in-the-muslim-world.html (Accessed 26 July 2011).

Dakss, Brian (2005). 'Pregnant Dog Adopts Hurt Squirrel', *CBS News*, 14th October [online] http://www.cbsnews.com/stories/2005/10/14/earlyshow/living/petplanet/main943873.shtml (Accessed 9th August 2011).

Darwin, Charles (1859). *On the Origin of Species by Means of Natural Selection, or the Preservation of Favored Races in the Struggle for Life*, London, John Murray.

Darwin, Charles (1868). *The Variation of Animals and Plants under Domestication*, London, John Murray.

Darwin, Charles (1871). *The Descent of Man, and Selection in Relation to Sex*, London, John Murray.

Dawkins, Richard (1976). *The Selfish Gene*, Oxford, UK, Oxford University Press.

Dawkins, Richard (1986). *The Blind Watchmaker*, New York, NY, Norton.

Dawkins, Richard (1995). *River Out of Eden: A Darwinian View of Life*, New York, NY, Basic Books.

Dawkins, Richard (1996). *Climbing Mount Improbable*, New York,

NY, Norton.

Dawkins, Richard (1998). 'Where do the real dangers of genetic engineering lie?', *The London Evening Standard*, 19 August.

Dawkins, Richard (2006). *The Selfish Gene (30th Anniversary edition)*. Oxford, United Kingdom, Oxford University Press.

Dawkins, Richard (2007). 'Genes still central', *New Scientist*, no. 2634 (15 December).

Depew, David J. (2010). 'Is Evolutionary Biology Infected With Invalid Teleological Reasoning?', Ann Arbor, MI, University of Michigan Library.

Dobzhansky, Theodosius (1967). *The Biology of Ultimate Concern*, New York, NY, New American Library.

Douglas, Angela E. (2010). *The Symbiotic Habit*, Princeton, NJ, Princeton University Press.

Eibl-Eibesfeldt, Irenäus (1955) 'Über Symbiosen, Parasitismus un andere besondere zwischenartliche Beziehungen tropischer Meeresfische'. *Zeitschrift für Tierpsychologie*, vol. 12, pp. 203-219.

Eibl-Eibesfeld, Irenäus (1970). *Love and Hate: The Natural History of Behavior Patterns*, trans. Strachan, Geoffrey (1971), Hawthorne, NY, Aldine de Gruyter.

Eldredge, Niles and Gould, Stephen Jay (1972). 'Punctuated equilibria: an alternative to phyletic gradualism', in Schopf, T.J.M. (ed.), *Models in Paleobiology*, San Francisco, Freeman Cooper, pp. 82-115.

Elias, Michael (1981) 'Serum cortisol, testosterone, and testosterone-binding globulin responses to competitive fighting in human males'. *Aggressive Behavior*, vol. 7, no. 3, pp. 215-224.

Everett, Hugh (1957). 'Relative State Formulation of Quantum Mechanics', *Reviews of Modern Physics*, vol 29, pp. 454-462.

Faber, Linda (2008). 'Kitty Corner: Update on an unlikely friendship', *The Sun Chronicle*, 21 July [online] http://www.thesunchronicle.com/articles/2008/08/13/pet_day/3396505.txt (Accessed 9 August 2011).

Falkowski, Paul G. (1997) 'Evolution of the nitrogen cycle and its influence on the biological sequestration of CO2 in the ocean', *Nature*, vol. 387, pp. 272-275.

Feder, Howard M. (1966). 'Cleaning Symbioses in the marine environment'. In Henry, S. Mark (ed.), *Symbiosis*, Vol. 1, pp. 327-280, New York, NY, Academic Press.

Feyerabend, Paul (1975). *Against Method: Outline of an Anarchistic Theory of Knowledge*, New York, NY, New Left Books.

Feyarabend, Paul (1988). *Farewell to Reason*, New York, NY, Verso.

Fryer, G. and Iles, T.D. (1972). *The Cichlid Fishes of the Great Lakes of Africa: Their Biology and Evolution*, Edinburgh, Oliver and Boyd.

Gadagkar, Raghavendra (2010), "Sociobiology in turmoil again", *Current Science*, vol. 99, no. 8, (25 October), pp. 1036-1041.

Geisler, Norman L. and Feinberg, Paul D. (1980). *Introduction to philosophy: a Christian perspective*, Grand Rapids, MI, Baker Book House.

Ghiselin, Michael (1974). *The Economy of Nature and the Evolution of Sex*, Berkley, CA, University of California Press.

Gilmore, Susan (2005). 'After goodbye visit, squirrel moves on', *The Seattle Times*, 27th December 2005, [online] http://community.seattletimes.nwsource.com/archive/?date=2 0051227&slug=squirrel27m (Accessed 6 August 2011).

Gimpl, Gerald and Fahrenholz, Falk (2001). 'The Oxytocin Receptor System: Structure, Function, and Regulation', *Physiological Reviews*, vol. 81 no. 2 (April), pp. 629-683.

Godfrey-Smith, Peter (2001). 'Three Kinds of Adaptationism', in Orzack, Steven H. and Sober, Elliott (eds.), *Adaptationism and Optimality*, Cambridge, UK, Cambridge University Press.

Gould, Stephen Jay and Lewontin, Richard C. (1979). 'The Spandrels of San Marco and the Panglossian Paradigm: a Critique of the Adaptationist Program'. *Proceedings of the Royal Society*, vol. 205, pp. 581-598.

Gould, Stephen Jay (1995). 'Ladders and Cones: Constraining

Evolution by Canonical Icons', in Silvers, Robert B (ed.), *Hidden Histories of Science*, New York, NY, New York Review of Books, pp. 37-68.

Gould, Stephen Jay (1997). 'Nonoverlapping Magisteria'. *Natural History*, no. 106, pp. 16-22.

Gould, Stephen Jay (1999). *Rocks of Ages*, New York, NY, Harmony Books, 1999.

Gould, Stephen Jay (2002). *The Structure of Evolutionary Theory*, Cambridge, MA, Belknap Press of Harvard University Press.

Griffiths, Paul E. and Gray, Russell D. (2001). 'Darwinism and Developmental Systems' in Oyama, Susan, Griffiths, Paul E. and Gray, Russell D., *Cycles of Contingency: Developmental Systems and Evolution*, Cambridge, MA, MIT Press.

Hamer, Dean H., Hu, Stella, Magnuson, Victoria L., Hu, Nan, Pattatucci, Angela M.L. (1993). 'A linkage between DNA markers on the X chromosome and male sexual orientation', *Science*, vol. 261, no. 5119 (July), pp. 321–7.

Harvard Magazine (2005), December, p. 33

Hatkoff, Isabella, Hatkoff, Craig, Kahumbu, Paula, Greste, Peter (2006). *Owen and Mzee: The True Story of a Remarkable Friendship*, New York, NY, Scholastic Press.

Hattenstone, Simon (2003). 'Darwin's child', *The Guardian*, 10 February.

Haught, John F. (1995). *Science and Religion: From Conflict to Conversation*, Mahway, New Jersey, Paulist Press.

Haught, John F. (2010). *Making Sense of Evolution: Darwin, God, and the Drama of Life*, Louisville, KY, Westminster John Knox Press.

Haught, John F. and Bateman, Chris (2011). 'Haught on Theology (1): Evolution vs. Religion', 16 August 2011 [online] http://onlyagame.typepad.com/only_a_game/2011/08/haught -on-theology-1-evolution-vs-religion.html (Accessed 16 August 2011).

Hayles, N. Katherine (1995). 'Narratives of Evolution and the

Evolution of Narratives', in Casti, J.L. and Karlqvist, Anders (eds.), *Cooperation and Conflict in General Evolutionary Processes*, New York, NY, John Wiley and Sons.

Hediger, Heini (1968). 'Putzer-fische im aquarium', *Natur und Museum*, vol. 98, pp. 89-96.

Henderson, M. (2001). 'Why you can't judge a man by his genes', *The Times*, 13 February.

Herre, Edward Allen, Machado, Carlos A, and West, Stuart A. (2001). 'Selective Regime and Fig Wasp Sex Ratios: Toward Sorting Rigor from Pseudo-Rigor in Tests of Adaptation', in Orzack, Steven Hecht and Sober, Elliott, *Adaptationism and Optimality*, Cambridge, UK, Cambridge University Press, pp. 191-218.

Hesse, Mary B. (1974). *The Structure of Scientific Inference*, Berkley, CA, University of California Press.

Heyd, David (1992). *Genethics: Moral Issues in the Creation of People*, Berkeley, CA, University of California Press.

Heyers D., Manns M., Luksch H., Güntürkün O. and Mouritsen H. (2007). 'A visual pathway links brain structures active during magnetic compass orientation in migratory birds'. *PLoS ONE* 2(9): e937.

Hobbes, Thomas (1651). *Leviathan, or the Matter, Forme, and Power of a Commonwealth, Ecclesiasticall and Civil*, London, Andrew Crooke.

Horizon: Nice Guys Finish First (1986). BBC, 4 April.

Hume, David (1739). *A Treatise of Human Nature*, Oxford, Clarendon Press.

Hume, David (1779). *Dialogues Concerning Natural Religion*, Edinburgh and London, William Blackwood and Sons.

Illich, Ivan (1971). *Deschooling Society*, London, Calder and Boyars Ltd.

Illich, Ivan (1975). *Medical Nemesis: The Expropriation of Health*, London, Marion Boyers.

Jensen, Christopher X. J. (2010). 'Robert Trivers and colleagues on

Nowak, Tarnita, and Wilson's "The evolution of eusociality"'
[online] http://www.christopherxjjensen.com/2010/10/13/
robert-trivers-and-colleagues-on-nowak-tarnita-and-wilsons-
the-evolution-of-eusociality/ (Accessed 22 June 2011).

Kant, Immanuel (1785). *Groundwork of the Metaphysics of Morals.*

Kant, Immanuel (1790). *Critique of Judgment*, translated by J. H.
Bernard (Mineola, NY: Dover Publications, 2005, original text
1790), p200.

Keller, Evelyn Fox (1992). *Secrets of Life, Secrets of Death: Essays on
Gender, Language and Science*, London and New York,
Routledge.

Kendrick, Keith M. (2004). 'The Neurobiology of Social Bonds',
Neuro-endocrinology Briefings, vol. 22, pp. 1-2.

Ketterson, E.D. and Nolan Jr, V. (1990). 'Site Attachment and Site
Fidelity in Migratory Birds: Experimental Evidence from the
Field and Analogies from Neurobiology;, in Gwinner, E. (ed.),
Bird Migration, Berlin, Springer-Verlag.

Kimura, Motō (1983). *The Neutral Theory of Molecular Evolution*,
Cambridge, UK, Cambridge University Press.

King, David (2002). 'The Human Genome Diversity Project',
GenEthics News, vol. 10.

Kluser, Stéphane; Neumann, Peter; Chauzat, Marie-Pierre; Pettis,
Jeffery S. (2010). 'Global honey bee colony disorders and
other threats to insect pollinators', *UNEP Emerging Issues*, pp.
1-12.

Kropotkin, Peter (1902). *Mutual Aid: A Factor of Evolution*,
London, William Heinemann.

Kuhn, Thomas S. (1962). *The Structure of Scientific Revolutions*
(Chicago, IL: University of Chicago Press, 1962), p138.

Lakoff, George and Johnson, Mark (1980). *Metaphors We Live By*,
Chicago, IL, University of Chicago Press.

Lee, Stan and Kirby, Jack (1963). *The X-Men*, #1 [Comic], New
York, Marvel Comics.

Lerner, Lawrence S. (2000). 'Good Science, Bad Science: Teaching

Evolution in the States', Washington, DC, The Thomas B Fordham Foundation.

Lewis, Eric B. (1978). 'A gene complex controlling segmentation in Drosophila'. *Nature*, vol. 276, pp. 565-570.

Lewontin, Richard C. (1978). 'Adaptation', *Scientific American*, vol. 239, pp. 212-228.

Lewontin, Richard C. (1991). *Biology as Ideology*, Toronto, ON, Anansi Press.

Lewontin, Richard C. (1995). 'Genes, Environment, and Organisms', in Silvers, Robert B (ed.), *Hidden Histories of Science*, New York, NY, New York Review of Books, pp. 115-140.

Lewontin, Richard C. (1997). 'Billions and Billions of Demons', *New York Review of Books*, [online] http://www.nybooks.com/articles/archives/1997/jan/09/billions-and-billions-of-demons/ (Accessed 24 June 2011).

Lewontin, Richard C. (2000). *It Ain't Necessarily So: The Dream of the Human Genome and Other Illusions*, New York, NY, New York Review Books.

Lewontin, Richard C. (2002). *The Triple Helix: Gene, Organism, and Environment*, Cambridge, MA, Harvard University Press.

Lovelock, James (1979). *Gaia: a New Look at Life on Earth*, Oxford, Oxford University Press.

Lowe, E. J. (1995). 'Metaphysics, Opposition to', in Honderich, Ted, *Oxford Encyclopedia of Philosophy*, Oxford, UK, Oxford University Press.

Lynch, Michael (2007). 'The frailty of adaptive hypotheses for the origins of organismal complexity', *Proceedings of the National Academy of Sciences of the United States of America*, vol. 104, supplemental. 1, pp. 8597-8604.

Mann, Charles (1994). 'Genes and Behavior', *Science*, vol. 264, no. 5166 (17 June 1994), pp. 1685-1739.

Margulis, Lynn (1970). *Origin of Eukaryotic Cells*, New Haven, CT, Yale University Press.

Margulis, Lynn (1991) 'Symbiogenesis and Sybionticism', in Margulis, Lynn and Fester, René (eds.), *Symbiosis as a Source of Evolutionary Innovation: Speciation and Morphogenesis*, Cambridge, MA, MIT Press.

Margulis, Lynn (1998). *Symbiotic Planet: A New Look at Evoluition*, New York, NY, Basic Books.

Maynard Smith, John (1964). 'Group selection and kin selection', *Nature* 201: 1145–1147.

Mendel, Gregor Johann (1866). 'Experiments in plant hybridization', trans. Druery, C.T and Bateson, William (1901), *Journal of the Royal Horticultural Society*, vol. 26, pp. 1–32.

Mereschkowski, Konstantin (1905). 'Über Natur und Ursprung der Chromatophoren im Pflanzenreiche'. *Biol. Centralbl.* vol. 25, pp. 593–604.

Michelson, Albert Abraham and Morley, Edward Williams (1887) "On the Relative Motion of the Earth and the Luminiferous Ether", *American Journal of Science*, no. 34 (1887): 333–345.

Midgley, Mary (1978). *Beast and Man: The Roots of Human Nature*, London and New York, Routledge.

Midgley, Mary (1985). *Evolution as a Religion: Strange hopes and stranger fears*, London and New York, Methuen.

Midgley, Mary (1991). *Can't We Make Moral Judgements?*, New York, NY, St. Martin's Press.

Midgley, Mary (1992). *Science as Salvation: A Modern Myth and its Meaning*, London and New York, Routledge.

Midgley, Mary (2003). *The Myths We Live By*, London and New York, Routledge.

Midgley, Mary (2011). 'Why The Idea Of Purpose Hasn't Gone Away', *Philosophy* (forthcoming).

Muir, William M. (1996). 'Group selection for adaptation to multiple-hen cages: selection program and direct responses'. *Poultry Science*, no. 75, pp. 447–458.

Muller, H. J. (1926). 'The Gene as the Basis of Life.' *Proceedings of* the *International Congress of Plant Sciences*, vol. 1, pp. 897-921.

Nagel, Ernest (1961). *The Structure of Science: Problems in the Logic of Scientific Explanation*. New York, NY, Harcourt, Brace and World.

National Academy of Sciences (2008). *Science, Evolution and Creationism* (Washington, DC: National Academies Press, 2008).

Neurath, Otto (1931). 'Physicalism: The Philosophy of the Vienna Circle'. In *Philosophical Papers*, edited by Robert S. Cohen and Marie Neurath, 48-51. Dordrecht: Reiderl, 1983.

Newman, John Henry (18730. *The Idea of a University*, London, Longmans, Green and Co.

Newman, Stuart A. (2006). 'The Developmental Genetic Toolkit and the Molecular Homology–Analogy Paradox', Biological Theory 1 (1), pp. 12-16.

Newman, Stuart A. (2009). 'Evolution Is Not Mainly a Matter of Genes', [online] http://www.councilforresponsiblegenetics org/projects/CurrentProject.aspx?projectId=9 (Accessed 21 June 2011).

Nietzsche, Friedrich (1887). *The Gay Science, 2nd Edition*, translated by Walter Kaufmann, (New York, NY: Vintage, 1974, original text 1887), section 344.

Nowak, Martin A., Tarnita, Corina E. and Wilson, Edward O. (2010). 'The evolution of eusociality', *Nature*, vol. 466, pp. 1057–1062.

Nowak, Martin A. and Sigmund, Karl (2007). 'How populations cohere: Five rules for cooperation' in May, Robert M., McLean, Angela. (eds.) *Theoretical Ecology: Principles and Applications*, Oxford, UK, Oxford University Press, pp. 7-16.

Nüsslein-Volhard, Christiane and Wieschaus Eric (1980). 'Mutations affecting segment number and polarity in Drosophila', *Nature* vol. 287, no. 5785, pp. 795–801.

Okasha, Samir (2004). Multilevel selection and the partitioning of covariance A a comparison of three approaches. *Evolution*, 58:

486–494.

Orr, H. Allen (1999). 'Gould on God: Can religion and science be happily reconciled?', Boston Review, October/November 1999, [online] http://bostonreview.net/BR24.5/orr.html (Accessed 22 June 2011).

Paley, William (1802). *Natural Theology; or, Evidences of the existence and attributes of the Deity, collected from the appearances of nature*, London, R. Faulder.

Parsons, Andrew (2001). 'Nature beats nurture', *The Times*, 13 February.

Passmore, John (1974). *Man's Responsibility for Nature: Ecological Problems and Western Traditions*, London, Duckworth.

Pellerin, Cheryl (1994). 'Ethical, Legal, and Social Issues of the Human Genome Project: What to Do with What We Know', *Environmental Health Perspectives*, vol. 102, no. 1 (January), pp. 58-59 [online] http://www.ncbi.nlm.nih.gov/pmc/articles/PMC1567247/pdf/envhper00389-0058-color.pdf (Accessed 22 June 2011).

Planck, Max (1949). *Scientific Autobiography and Other Papers*, translated by F. Gaynor (New York, NY: 1949, original text 1948), pp 33-34.

Plato (circa 360 BC). *Timaeus*.

Popper, Karl (1959). *The Logic of Scientific Discovery*, New York, NY, Basic Books,

Popper, Karl (1976). *Unended Quest; An Intellectual Autobiography*, La Salle, IL, Open Court.

Random House Unabridged Dictionary (2011). "sacred", [online] http://dictionary.reference.com/browse/sacred (Accessed 2 January 2011).

Reiss, John O. (2005). 'Natural Selection and the Conditions for Existence: Representational vs. Conditional Teleology in Biological Explanation', *History and Philosophy of the Life Sciences*, vol 27, pp. 249-280.

Reiss, John O. (2009). *Not By Design: Retiring Darwin's*

Watchmaker, University of California Press, 2009.

Reiss, John O. and Bateman, Chris (2011). 'Reiss against Evolutionary Design', *Only a Game*, 2 August 2011 [online] http://onlyagame.typepad.com/only_a_game/2011/08/reiss-against-adaptation.html (Accessed 2 August 2011).

Roberts, Mary (1834). *The Conchologists Companion*, London, Whittaker and Co.

Rose, Kenneth D. (2006). *The Beginning of the Age of Mammals*, Baltimore, MD, Johns Hopkins University Press.

Rosenberg, Alexander and Bouchard, Frederic (2008). 'Fitness' in Zalta, Edward N. (ed.), *The Stanford Encyclopedia of Philosophy (Fall 2010 Edition)*, available online http://plato.stanford.edu/archives/fall2010/entries/fitness/, retrieved 20 June 2011.

Rudwick, Martin J. S. (1964). 'The inference of function from structure in fossils', *British Journal for the Philosophy of Science*, no. 15, pp. 27-40.

Ruse, Michael (2003a). 'Is Evolution a Secular Religion?', *Science*, vol. 299, no. 5612, pp. 1523-1524.

Ruse, Michael (2003b). *Darwin and Design: Does Evolution Have a Purpose?*, Cambridge, MA, Harvard University Press.

Ruse, Michael (2010). *Science and Spirituality: Making Room for Faith in the Age of Science*, Cambridge, MA, Cambridge University Press.

Russell, Bertrand (1946). *A History of Western Philosophy*, London, George Allen and Unwin.

Sagan, Lynn (1967). 'On the origin of mitosing cells.' Journal of Theoretical Biology, vol. 14, no. 3, pp. 255-274.

Segal, Charles (1986). *Interpreting Greek tragedy: myth, poetry, text*, Ithaca, NY, Cornell University Press.

Schrödinger, Erwin (1948). *What is Life?*, Cambridge, UK, Cambridge University Press.

Shalchian-Tabrizi, Kamran; Minge, Marianne A.; Espelund, Mari; Orr, Russel; Ruden, Torgeir; Jakobsen, Kjetill S.; Cavalier-Smith, Thomas (2008). 'Multigene phylogeny of choanozoa and the origin of animals'. *PLoS ONE*, vol. 3, no. 5, e2098 [online] http://www.plosone.org/article/info%3Adoi%2F10 .1371%2Fjournal.pone.0002098 (Accessed 3 August 2011).

Singer, Bryan (2000). *X-Men* [Film], Hollywood, Paramount Pictures.

Sober, Elliott (1988). 'What Is Evolutionary Altruism?', *Canadian Journal of Philosophy*, supplementary vol. 14, p.75-99.

Sober, Elliott (2000a). 'Two faces of fitness', in Singh, R.S. et al. (eds.), *Thinking About Evolution: Historical, Philosophical, and Political Perspectives*, Cambridge, UK, Cambridge University Press.

Sober, Elliott (2000b). 'Psychological Egoism,' in LaFollette, Hugh (ed.), *The Blackwell Guide to Ethical Theory*, Oxford, UK, Blackwell.

Sober, Elliott, and Wilson, David Sloan (2011), *"Adaptation and Natural Selection* Revisited", Journal of Evolutionary Biology, 24: 462–468

Smolin, Lee Smolin (1997). *The Life of the Cosmos*, New York, NY, Oxford University Press.

Swenson, William, Wilson, David Sloan. and Elias, Robert (2000). 'Artificial Ecosystem Selection'. *Proceedings of the National Academy of Sciences of the United States of America*, vol. 97, pp. 9110-9114.

Taylor, Charles (2007). *A Secular Age*. Cambridge, MA, Belknap of Harvard.

Tennekes, Henk (2010). *The Systemic Insecticides: A Disaster in the Making*, Zutphen, Netherlands, Weevers Walburg Communicatie.

Thompson, Nicholas S. (2000). 'Shifting the Natural Selection Metaphor to the Group Level', *Behavior and Philosophy*, vol. 28, pp. 83-101.

Trivers, Robert L. (1971). 'The Evolution of Reciprocal Altruism'. *The Quarterly Review of Biology*, vol. 46, no. 1, pp. 35–57.

van Leengoed, Eric, Kerker, E, Swanson, Heidi H. (1987).

'Inhibition of post-partum maternal behaviour in the rat by injecting an oxytocin antagonist into the cerebral ventricles.' *Journal of Endocrinology*, vol. 112, no. 2 (February), pp. 275-82.

Venkatesh, Byrappa; Si-Hoe, San Ling; Murphy, David and Brenner, Sydney (1997). 'Transgenic rats reveal functional conservation of regulatory controls between the Fugu isotocin and rat oxytocin genes', *Proceedings of the National Academy of Sciences of the United States of America*, vol. 94, no. 23 (November), pp. 12462-12466.

Wagner, Andreas (2008). 'Neutralism and selectionism: a network-based reconciliation', *Nature Reviews Genetics*, vol. 9 (December), pp. 965-974.

Wallin, Ivan Emanuel (1923). 'The Mitochondria Problem'. *The American Naturalist*, vol. 57, no. 650, pp. 255–61.

Watson, James (2004). *DNA: The Secret of Life*, London, Arrow Books.

Walton, Kendall L. (1990). *Mimesis as Make-believe: On the Foundations of the Representational Arts*, Cambridge, MA, Harvard University Press.

Weigmann, Katrin (2004). 'The code, the text and the language of God', *EMBO Reports*, no. 5, pp. 116-118, [online] http://www.nature.com/embor/journal/v5/n2/full/7400069.html (Accessed 28 June 2011).

White, Jr., Lynn Townsend (1967). 'The Historical Roots of Our Ecological Crisis', *Science*, vol. 155, no. 3767, March 10, pp. 1203–1207.

Wigmore, Barry (2007). 'The tiger who adopted a litter of piglets (but is it a tale full of porkies?)', *Daily Mail Online*, 30 November, [online] http://www.dailymail.co.uk/news/article-498789/The-tiger-adopted-litter-piglets-tale-porkies.html (Accessed 9 August 2011).

Williams, George C. (1966). *Adaptation and Natural Selection.* Princeton University Press, Princeton, N.J.

Willis, Arthur J. (1997), 'The ecosystem: an evolving concept viewed historically', *Functional Ecology*, vol. 11, no. 2, pp. 268-271.

Wilson, Edward O. (1975). *Sociobiology: The New Synthesis*, Cambridge, MA, Harvard University Press.

Wilson, Edward O. (1992). *The Diversity of Life*, Cambridge, MA, Harvard University Press.

Wilson, Edward O. (1998). *Consilience: The Unity of Knowledge*, New York, NY, Vintage.

Wilson, Edward O. (2008). 'One giant leap: How insects achieved altruism and colonial life'. BioScience. 58 (1): 17-25.

Wilson, David Sloan (1975). 'A theory of group selection'. *Proceedings of the National Academy of Sciences of the United States of America*, vol. 72, pp. 143–146.

Wilson, David Sloan (1979). *The Natural Selection of Populations and Communities*, Menlo Park, CA, Benjamin Cummings.

Wilson, David Sloan (2009a). 'V: The Patriotic History of Individual Selection Theory' [online] http://scienceblogs.com/evolution/2009/10/truth_and_reconciliation_for_g_3.php (Accessed 22 June 2011).

Wilson, David Sloan (2009b). 'XI: Dawkins Protests (Too Much)' [online] http://scienceblogs.com/evolution/2009/11/truth_and_reconciliation_for_g_9.php (Accessed 22 June 2011).

Wilson, David Sloan (2009c). 'XIII: Hamilton Speaks' [online] http://scienceblogs.com/evolution/2009/11/truth_and_reconciliation_for_g_11.php (Accessed 22 June 2011).

Wise, Robert (1979). *Star Trek: The Motion Picture* [Film], Hollywood, Paramount Pictures.

Wittgenstein, Ludwig (1962). *Philosophical Investigations*, translated by Anscombe, G.E.M., New York, NY, Macmillan.

Wolf, Gary (2011). 'The Church of the Non-Believers', *Wired*, vol. 14, no. 11 (November 2006), available online http://www.wired.com/wired/archive/14.11/atheism.html. Retrieved 2nd January 2011.

Wolfram, Stephen (2002). *A New Kind of Science*, Champaign, IL,

Wolfram Media.

Yablo, Stephen (1998). 'Does Ontology Rest on a Mistake?', *Proceedings of the Aristotelian Society*, Supplementary vol. 72, no. 1, pp. 229–262.

Young, Anne (2010). 'The monkey and the kitten', The Guardian, 25 August, [online] http://www.guardian.co.uk/environment /gallery/2010/aug/25/bali-animals#/?picture=366113722 &index=0 (Accessed 9 August 2011).

Zumbach, Clark (1984), *The transcendent science: Kant's conception of biological methodology*, The Hague and Boston and Hingham, MA, M. Nijhoff.

Zylinska, Joanna (2009). *Bioethics in the Age of New Media*, Cambridge, MA, MIT Press.

Zylinska, Joanna and Bateman, Chris (2009). 'Joanna Zylinska on Bioethics',   [online] http://onlyagame.typepad.com/only_a_game/2009/09/joanna-zylinska-on-bioethics.html (Accessed 28 June 2011).

Glossary

Absolute truth: the belief that there is only one coherent account of the truth, that it is objective and knowable, and that because there can be only one such accurate account any claim that conflicts with it is by definition false. Compare *perspectival truth* (q.v.).

Adaptation: [1] any feature of an organism that demonstrates a high degree of conformity to its environment e.g. the fins of a fish are adapted to life in water. [2] the process by which this happens i.e. *natural selection* (q.v.) in response to environmental circumstances.

Adaptationism: a belief that *adaptation* (q.v.) is a phenomena of special significance either because it can fully account for the outcomes of evolutionary processes, or because it solves the most significant problems in evolutionary biology.

Advantages persist: an alternative myth to the *selfish gene* (q.v.) that suggests biological and behavioral benefits tend to persevere precisely because they are advantageous.

Allele: a particular variation of a *gene* (q.v.) e.g. if eye color was caused by a single gene, there would be an allele for blue eyes and an allele for brown eyes. This book uses *gene variant* as a simplified synonym for allele.

Artificial selection: *selection* (q.v.) that occurs as part of a controlled experiment or breeding program.

Atheology: the examination of religious issues from a rational perspective, based on a rejection of any viable concept of God. Compare *theology* (q.v.).

Authorized story: in *pretence theory* (q.v.), any fiction that is accepted as fact can be considered an authorized story.

Base pair: a set of two molecules that bond to each other to make up the individual segments in *DNA* or *RNA* (q.q.v.).

Cambrian explosion: the appearance approximately 530 million years ago of almost all major animal forms over an interval of a few million years.

Chain of inheritance: an alternative myth to the *ladder of progress* (q.v.) which looks from the present to the past and sees contemporary biology as connected to ancestral biology by an unbroken sequence of descendents.

Conditions for existence: Cuvier's idea that all the features of an animal must match its lifestyle e.g. an animal with teeth suited for meat must have a digestive system suited for meat. Also. an alternative myth to *adaptationism* (q.v.).

Conserved: referring to a *gene* (q.v.) that persists for very long periods of time, and thus can be found in the DNA of many different species.

Co-operation is an advantage: an alternative myth to *kin selection* (q.v.) recognizing that co-operation between and within species offers survival benefits irrespective of whether the creatures in question are related. See also *symbiosis, trust is an advantage* (q.q.v.).

Dangerous belief: three different religious or non-religious excesses of belief, namely *superstition, enthusiasm,* and *fanaticism* (q.q.v.), based on terms used in the late eighteenth century.

Design argument: an argument for the existence of God, proceeding by inferences from the presence of order or complexity in the Universe, through to a claim of purpose or design, and hence to a deity.

DNA: deoxyribonucleic acid, a long, double helix chain molecule, essential to life as we know it, and comprising of a great many *base pairs,* forming both *genes* and *non-coding sections* (q.q.v.).

Ecosystem: a biological environment, including all the living creatures and the non-living elements they depend upon e.g. air, soil, rocks, earth, sunlight.

Egoism: a theory about motivation claiming that we act only in our self-interest, or that our ultimate desires are self-directed (also called: psychological egoism). See also *hedonism* (q.v.).

Enthusiasm: for religions, the certainty that one has heard the voice of God; for *non-religions*, strident certainty, especially concerning *atheology* (q.q.v.). One of three kinds of *dangerous belief* (q.v.). See also *superstition* and *fanaticism* (q.q.v.).

Evolution: life changing over time, for whatever reason.

Extinction by lottery: Gould's concept that the vast majority of species that become extinct are not out-competed by rival creatures but die out essentially by chance.

Fanaticism: for both religions and *non-religions* (q.v.), the kind of certainty that licenses going beyond the common moral order. One of three kinds of *dangerous belief* (q.v.). See also *superstition* and *enthusiasm* (q.q.v.).

Fictional: any circumstance that is imagined can be considered fictional. The term is also a synonym for 'fictionally true' in Walton's *pretence theory* (q.v.).

Fitness: an imaginary measure of an organism's capacity to survive and reproduce. See also *fitness to environment, fitness as propensity* and *genetic fitness* (q.q.v.).

Fitness to environment: Darwin's metaphorical idea that different organisms 'fit' their environment better than others, thus allowing them to be 'selected'. See also *selection* (q.v.).

Fitness as propensity: a concept of *fitness* (q.v.) equated to expected numbers of offspring by various means.

Functional chain: a sequence of *RNA* encoded by a particular *gene* (q.q.v.) that serves a particular biological function, such as modifying the activity of other genes.

Gaia principle: a view of life that imagines the entirety of the living world and its inorganic surroundings as a single self-regulating complex system, and the scientific theory that this system plays a crucial role in maintaining an environment capable of supporting life.

Gene: a section of *DNA* that encodes a specific *protein* or a *functional chain* (q.q.v). Genes are units of heredity that occupy specific positions in an organism's *genome* (q.v.).

Gene pattern: an informal name for *genotype* (q.v.), used in this book in order to simplify the terminology.

Gene supremacy: the belief that the *gene-centered view* (q.v.) provides the most effective explanation for behavior in all cases.

Gene variant: an informal name for *allele* (q.v.), used in this book to simplify the terminology.

Gene's eye view: another name for the *gene-centered view* (q.v.).

Gene-centered view: an imaginary metaphorical perspective in which *genes* (q.v.) are imagined to be causative agents, aiming to increase the rate of incidence of their particular pattern of information.

Genetic determinism: in its strongest form, the claim that *genes* (q.v.) determine behavior. More commonly, the weaker claim that genes strongly influence behavior, or that behavior is determined by the confluence between genetics and environment.

Genetic drift: the change in the frequency of a *gene variant* in a *population* (q.q.v.) as a result of random sampling.

Genetic fitness: *fitness* considered in terms of change in gene ratios i.e. in terms of the number of instances of its *genes* (q.q.v.) an organism is likely to pass into the next generation.

Genetic innovation: see *theory of genetic innovation* (q.v.).

Genetic toolbox: a term used in this book to refer to the developmental-genetic toolkit, the set of *genes* (q.v.) that determine the body plan of animals i.e. how many legs and arms, and how everything is connected.

Genome: the entire chemically-encoded hereditary information for a particular organism i.e. all *genes* and *non-coding sections* (q.q.v).

Genotype: the specific genetic make-up of a particular individual cell or animal. This book uses *gene pattern* as a simplified synonym for genotype.

Grand narrative: Hayles term for the stories a culture tells in order to make sense of the world. Compare Campbell's *living mythology* (q.v.).

Group selection: *selection* (q.v.) at a level of abstraction beyond the individual. See also *multi-level selection* (q.v.).

Haldane's jest: *inclusive fitness* (q.v.) stated in the form of a joke about how many relatives an animal will be willing to die for, based on the sum of their fractional relatedness e.g. (in Haldane's original version) "I am willing to die for four uncles or eight cousins!"

Hamilton's rule: an optimality equation in *kin selection* (q.v.), derived by W.D. Hamilton, whereby a gene encouraging self-sacrifice can propagate if the benefits to the recipients multiplied by their relatedness to the sacrificing animal exceeds the costs. See also: *inclusive fitness* (q.v.).

Hedonism: a theory about motivation claiming that our ultimate desires are only concerned with seeking pleasure and avoiding pain (also called: psychological hedonism). See also *egoism* (q.v.).

How-Why game: a kind of *teleological* claim that purports to explain an *adaptation* (q.q.v.) by means of an explanatory cause, but in fact simply presumes the cause suffices as an explanation and proceeds to invent a story consistent with what is observed.

Hox genes: the components of the *genetic toolbox* (q.v.).

Inclusive fitness: an imaginary extension of the metaphorical concept of *genetic fitness* in which the *gene variants* (q.q.v.) passed on via offspring are numerically supplemented by the equivalent gene variants passed on by relatives modified by a fractional value representing relatedness.

Individual self-interest: see *egoism* (q.v.).

Information: any sequence of symbols. Also, the mythic view that various phenomena can be understood in terms of such sequences.

Intelligent design: assertion of the *design argument* as an explanation for *adaptation* (q.q.v). Also, Intelligent Design (in capitals), a political movement advocating a theologically neutral version of the design argument.

Kin selection: *selection* (q.v.) for behaviors that favor an animal's genetic relatives over the animal itself. Also: in this book, the myth that all co-operation is self-interested because the only kinds of co-operation that don't afford mutual benefit (allegedly) occur only in respect of relatives. Compare *co-operation is an advantage* and *trust is an advantage* (q.q.v).

Ladder of progress: the mythic claim that the history of the evolution of life shows inherent signs of increasing complexity. Compare *chain of inheritance* (q.v.).

Living mythology: Campbell's term for a set of stories that serve an active role in a particular culture, shaping the ethics and norms of that society.

Logical positivism: the *positivist* movement of the Vienna Circle in the 1920s and 1930s that considered *metaphysics* (q.q.v) as meaningless because they could not be verified.

Mega-organism: the metaphorical view that all life on Earth can be understood *as if* it were a single *organism* (q.v.). See also *Gaia principle* (q.v.).

Megatext: Charles Segal's term for collections of myths or stories that imply a single fictional world.

Metaphysics: the philosophical exploration of that which cannot be tested or proven i.e. claims concerning reality that exceed experience or direct evidence.

Metaphysical: the adjective conjugated from *metaphysics* (q.v.).

Metaphor of design: the imaginary explication of an *adaptation* (q.v.) *as if* it had been designed, whether or not a designer is additionally postulated. Also, an alternative myth to *intelligent design* (q.v.). See also *design argument* (q.v.).

Migration: the seasonal travelling of animals of various kinds from one territory to another.

Multi-level selection: a contemporary interpretation of *group selection*, whereby the effects of *selection* can be identified at different scales e.g. individuals, packs and families, *trait groups, ecosystems* (q.q.v.) etc.

Mutation: a change in the *DNA* for a particular *gene* that produces a new *gene variant* (q.q.v.).

Myth: an imaginative pattern, especially one that is fictional or metaphorical but that is not usually noticed as such.

Natural selection: differential survivorship, i.e. the kind of *selection* that occurs in nature, as opposed to *artificial selection* (q.q.v).

Natural theology: reasoning from observation of the world to claims about God or gods. The field includes the *design argument*. Compare *theology of nature* (q.v.).

Neutral mutation: in the *neutral theory*, a *mutation* that does not significantly affect the biological function that a *gene* (q.q.v.) contributes toward.

Neutral theory: Kimura's idea that the vast majority of *mutations* (q.v.) don't have a significant effect on the animal in question.

Non-coding section: portions of genetic material that don't specify or regulate the production of *proteins* (q.v.), sometimes called 'junk DNA'.

Non-religion: a belief-system with its own mythology and values that is not a religious tradition e.g. Marxism, *positivism* (q.v.).

Normal science: Kuhn's term for periods of history during which a particular scientific field operates within a particular *paradigm* (q.v.).

Nova effect: Charles Taylor's term for the ideological fragmentation in the modern world within which the range of possible beliefs tends towards infinite diversity.

Organism: a living creature, irrespective of whether it is single celled or multi-cellular. In the case of multi-cellular organisms, the term implies the metaphorical view that the collection of cells can be treated as one living thing rather than many. See also *super-organism* and *mega-organism* (q.q.v.).

Orthodox science fiction: any fictional story that does not overtly contradict a contemporary perspective on the *science megatext* (q.v.).

Paradigm: Kuhn's term for the collection of symbolic generalizations, experimental methods and common assumptions shared by scientists in any given field and historical era.

Perspectival truth: the belief that the true state of affairs can be attained only by examination from multiple different perspectives, many of which may contribute usefully to determining what is true. Compare *absolute truth* (q.v.).

Phenotype: an organism's characteristics, specifically in relation to its *genes* i.e. its *genotype* (q.q.v.). This book avoids using this term, preferring to talk about animals directly whenever possible.

Physicalism: the belief that only physical matter exists, also known as materialism.

Physicalist: a person who believes in *physicalism* (q.v.).

Popper's milestone: a metaphorical boundary between (testable) science and (untestable) *metaphysics* (q.v.), based on the work of Karl Popper.

Population: a group of animals that interbreed.

Positivism: a scientific *non-religion* based upon avoiding belief in untestable things, and thus an attempt to minimize *metaphysics* (q.q.v.) as much as possible.

Positivist: a person who believes in some form of *positivism* (q.v.).

Pretence theory: the part of Walton's philosophical make-believe theory of representation concerning the relationship between fiction and fact.

Propensity interpretation: see *fitness as propensity*.

Protein: a molecule which forms the basic building block of life, the chemical structure of which is defined by the sequence of *base pairs* in one or more *genes* (q.q.v.).

Punctuated equilibrium: Eldredge and Gould's theory that evolutionary developments occur primarily in isolated periods of rapid change between extended periods of stasis, when there is little or no change.

Realm of ends: Kant's mythic image of communal autonomy, whereby each person works towards the fulfillment of their own ideals and goals, while avoiding blocking other people from attaining their own objectives wherever possible.

Refinement of possibilities: an alternative myth to *survival of the fittest* (q.v.) that stresses the manner in which successive generations of creatures have gained refined capabilities by overcoming the survival challenges they faced.

RNA: ribonucleic acid, a long, single-stranded chain molecule essential to life as we know it. RNA chains are transcribed from *DNA* (q.v.), or copied from other RNA chains.

Science as truth: the *metaphysical* (q.v.) belief that science necessarily progresses towards a more accurate and truthful description of the world. Compare *truth in fiction* (q.v.).

Science megatext: the collection of texts (e.g. experiments, theories) that comprise the current *paradigm* for scientific research, acting as a *megatext* (q.q.v.).

Scientific revolution: Kuhn's term for periods of history during which a specific scientific *paradigm* (q.v.). is overthrown and replaced.

Selection: a metaphorical term, originally used by Darwin, describing how environmental circumstances favor some animals over others, allowing them to reproduce more successfully. An animal that succeeds in this way is said to have been 'selected'.

Self-interest: see *egoism* (q.v.).

Selfish gene: a metaphor for the *gene-centered view* that also implies *egoism* and *gene supremacy* (q.q.v.). Compare *advantages persist* (q.v.).

Sexual selection: Darwin's idea that animals conduct *selection* on their potential mates, and that this is in addition to the differential survivorship implied by *natural selection* (q.q.v.).

Spandrel: Gould and Lewontin's term for a biological feature that is not an *adaptation*, that is, any aspect of an organism that was not the product of direct *natural selection* (q.q.v.).

Spencerism: belief in the political, economic or biological necessity of competition, as embedded in the myth of *survival of the fittest* (q.v.), coined by Herbert Spencer.

Super-organism: the metaphorical view that a collection of different multi-cellular *organisms* (q.v.) that operate in a closely integrated fashion can be understood *as if* they were a single organism e.g. an ant nest or a Portuguese Man o' War.

Superstition: for religions, faith in magic; for *non-religions* (q.v.), belief in outlandish science fiction fantasies. One of three kinds of *dangerous belief* (q.v.). See also *enthusiasm* and *fanaticism* (q.q.v.).

Survival of the fittest: Spencer's economic concept of success necessarily proving superiority that was later shoehorned into Darwin's theory. Compare *refinement of possibilities* (q.v.).

Symbiosis: persistent co-operation between organisms who are not related.

Teleological: exhibiting or implying purpose, or directed towards particular outcomes, particularly purpose-focused explanations of circumstances.

Teleology: see *teleological* (q.v.).

Theology: the examination of religious issues from a rational perspective, in relation to a specific concept or set of concepts of God. Compare *atheology* (q.v.).

Theology of nature: arguing for reinterpretation of religious beliefs in the light of scientific conclusions about the nature of life. Compare *natural theology* (q.v.).

Theory of genetic innovation: Andreas Wagner's proposal that advantageous *genes* come about through an initial period of the accumulation of *neutral mutations*, followed (after a *mutation* creates a new genetic advantage) by a period of *selection* (q.q.v).

Toolbox genes: an informal name for *hox genes* (q.v.). See also *genetic toolbox* (q.v).

Trait-group: David Sloan Wilson's term for a set of organisms in which the actions of any individual within the group affect every individual in the group. Note that the individuals *need not* be related. See also *multi-level selection* (q.v.).

Trust is an advantage: an alternative myth to *kin selection* (q.v.) that recognizes the survival benefits offered by sociality and trust. See also *co-operation is an advantage* (q.v.).

Truth in fiction: an alternative myth to *science as truth* (q.v.) suggesting that rather than fiction and fact being polar opposites, facts are a particular kind of fiction.

Watchmaker analogy: Paley's idea that if we found a watch, we would imply a watchmaker, and that analogous reasoning can be used to defend the *design argument* (q.v.).

Williams Principle: the idea, suggested by George C. Williams that *adaptation* at a level requires *selection* (q.q.v.) at that level e.g. group adaptations must be the result of selection at the level of groups. See also *multi-level selection* (q.v.).

World's eye view: another name for *Gaia principle* (q.v.).

Contemporary culture has eliminated both the concept of the public and the figure of the intellectual. Former public spaces – both physical and cultural – are now either derelict or colonized by advertising. A cretinous anti-intellectualism presides, cheerled by expensively educated hacks in the pay of multinational corporations who reassure their bored readers that there is no need to rouse themselves from their interpassive stupor. The informal censorship internalized and propagated by the cultural workers of late capitalism generates a banal conformity that the propaganda chiefs of Stalinism could only ever have dreamt of imposing. Zer0 Books knows that another kind of discourse – intellectual without being academic, popular without being populist – is not only possible: it is already flourishing, in the regions beyond the striplit malls of so-called mass media and the neurotically bureaucratic halls of the academy. Zer0 is committed to the idea of publishing as a making public of the intellectual. It is convinced that in the unthinking, blandly consensual culture in which we live, critical and engaged theoretical reflection is more important than ever before.